REVISE EDEXCEL GCSE
Statistics

REVISION WORKBOOK

Series Consultant: Harry Smith

Author: Navtej Marwaha

A note from the publisher

In order to ensure that this resource offers high-quality support for the associated Pearson qualification, it has been through a review process by the awarding body. This process confirms that this resource fully covers the teaching and learning content of the specification or part of a specification at which it is aimed. It also confirms that it demonstrates an appropriate balance between the development of subject skills, knowledge and understanding, in addition to preparation for assessment.

Endorsement does not cover any guidance on assessment activities or processes (e.g. practice questions or advice on how to answer assessment questions), included in the resource nor does it prescribe any particular approach to the teaching or delivery of a related course.

While the publishers have made every attempt to ensure that advice on the qualification and its assessment is accurate, the official specification and associated assessment guidance materials are the only authoritative source of information and should always be referred to for definitive guidance.

Pearson examiners have not contributed to any sections in this resource relevant to examination papers for which they have responsibility.

Examiners will not use endorsed resources as a source of material for any assessment set by Pearson.

Endorsement of a resource does not mean that the resource is required to achieve this Pearson qualification, nor does it mean that it is the only suitable material available to support the qualification, and any resource lists produced by the awarding body shall include this and other appropriate resources.

For the full range of Pearson revision titles across GCSE, BTEC and AS/A Level visit:
www.pearsonschools.co.uk/revise

Contents

1-to-1 page match with the Statistics Revision Guide ISBN 9781292098296

....

A small bit of small print

Edexcel publishes Sample Assessment Material and the Specification on its website. This is the official content and this book should be used in conjunction with it. The questions have been written to help you practise every topic in the book.

Remember: the real exam questions may not look like this.

Types of data

1 Which of these describe qualitative information? Put a tick in the correct boxes.

(a) The number of children in a family	
(b) The football team a person supports	
(c) The colour of a person's hair	
(d) The cleverness of a person	
(e) The position of an athlete in a race	

(2)

2 Which of these describe ranked data? Put a tick in the correct boxes.

(a) The number of children in a family	
(b) The position a person comes in a test	
(c) The mark a person got in an exam	
(d) The position, by age, in a family	
(e) The grade in an exam	

(2)

3 Is each of the following qualitative or quantitative data?

(a) The political party a person supports ....

> The political party a person supports is qualitative data because it has to be described using words.

(b) The number of children crossing a road each day ....

(c) The types of different plants in a garden ....

(d) The loudness of a television .... **(4)**

4 Is each of the following discrete or continuous data?

(a) The position an athlete comes in a race ....

(b) The size of a dress ....

(c) The weight of a fish .... **(3)**

5 Here is a list of words used in statistics.

quantitative   discrete   qualitative   rank   continuous

Complete each of the sentences using a word from the list.

(a) A .... variable is one which cannot be given a numerical value.

(b) The number of people on a bus is an example of a .... variable

which is ....

(c) An example of a .... is the position that someone comes in a test. **(3)**

6 Magda wants to do a survey of the eye colour of people in her class.
State a problem she might have recording the data.

....

.... **(1)**

Measurements and variables

> **Guided**   **1**   Each statement below describes a pair of observations.

Complete each statement using 'response' or 'explanatory' for the variables.

(a) The number of flowers in a garden is the _response_ variable and the amount of rainfall

is the _explanatory_ variable.

> A change in rainfall can cause more flowers
> to grow but not the other way round.

(b) The length of a child's arm is the .... variable and their age is the

.... variable.

(c) The price of a house is the .... variable and the number of rooms is the

.... variable.

(d) The number of hours of revision is the .... variable and the grade in the

exam is the .... variable.   **(4)**

2   For each of the following state:

(i) the smallest possible value it could be

(ii) the largest possible value it could be.

(a) The height of a plant is 24 cm to the nearest cm.   (i) .... (ii) ....

> The unit of measurement is cm. To get the smallest
> possible height take off 0.5 cm.
> To get the largest possible height add on 0.5 cm.

(b) The weight of a cow is 242 kg to the nearest kg.   (i) .... (ii) ....

(c) The time taken to complete an essay is 2 hours to the nearest minute.

    (i) .... (ii) ....   **(3)**

3   John timed how long, in seconds correct to the nearest second, it took him to read ten
different text messages, all of different lengths.

Complete these sentences.

(a) The length of time it took John to read a text message is an example of a quantitative

variable which is ....

(b) The number of characters in a text message is an example of a quantitative variable

which is ....

(c) The explanatory (independent) variable is ....

(d) The response (dependent) variable is ....

John timed one of the text messages at 19 seconds, correct to the nearest second.

(e) What is the shortest time it could be?   ....   **(5)**

Sampling frames, pre-tests and pilots

1 Explain why each of the following sampling methods is biased.

    (a) A manager of a sports centre wants to know how frequently people do sports. He asks the first 20 people who arrive on Monday morning.

    ....

    (b) A teacher wants to know about the average number of children in a family. He uses the numbers of children in the families of the children in his class.

    ....

    (c) A market researcher takes a sample based on the telephone directory.

    ....

    (d) Uzma wants to find out about the use of the town library. She asks people at the library in the evening whether they would use it more if it was open on Sundays.

    ....

    (e) A company wants to know whether people are satisfied with their purchases. They contact everyone for whom they have an email address.

    .... **(5)**

Guided

2 Suggest a suitable sampling frame for each of the following samples.

    (a) A head teacher wants to know whether students like the idea of a year 11 prom.

    A list of all year 11 students   [The sampling frame must contain all members of the population.]

    (b) A garage manager wants to know whether all the repairs done last month were satisfactory.

    ....

    (c) A local councillor wants to find out the opinions of householders on an estate.

    ....

    (d) The environmental services department wants to find out what people think of the waste collection services in a village.

    .... **(4)**

3 Akbar wants to investigate the proportion of people in his town who eat organic foods.
He plans to ask people questions as they leave the local organic food shop.
Explain why this is not a good place to ask people.

    .... **(1)**

4 (a) Write down **one** advantage and **one** disadvantage of a census.

    .... **(1)**

    (b) Write down **one** advantage and **one** disadvantage of a sample.

    .... **(1)**

5 Here is a list of random numbers.   [Don't forget that the numbers must be between 0 and 49.]

    61 38 41 73 03 90 59 79 68 80 73 72 60 48 91 34 81 92
    07 84 81 29 94 70 94 69 13 31 72 10 39 47 90 00 85 04
    Starting with the number 38, write down 10 random numbers between 0 and 49.

    .... **(1)**

Experiments and hypotheses

1 Andrew thinks that there might be a relationship between the number of different types of vegetables people grow in their gardens and the size of the garden.

Write down a suitable hypothesis for Andrew.

.... **(1)**

> A hypothesis has to be phrased as a statement **not** as a question.

2 Jenny thinks that adding salt to chips makes them taste better.

Write down a suitable hypothesis for Jenny.

.... **(1)**

Guided **3** In each case describe what control would be used to investigate the hypothesis.

(a) When seeds are not watered, they will not germinate and start to grow.

.... **(1)**

(b) Adding copper sulphate to plants' water makes their flowers turn blue.

Observe the flower colour of a group of similar plants watered with distilled water. **(1)**

> Not adding copper sulphate is a control to see if the flowers turn blue anyway.

(c) Drinking coffee before bedtime keeps people awake.

.... **(1)**

4 Tom wants to find out how effective slug pellets are in getting rid of slugs.

(a) Write down a suitable hypothesis for Tom.

.... **(1)**

(b) Describe an experiment Tom could carry out to test his hypothesis.
You must include details of a control.

....

.... **(2)**

Guided **5** There are 250 students in year 11 of a school.

24 students out of a sample of 40 said that they could cycle to school.

Work out an estimate for the number of students in year 11 who could cycle to school.

Proportion of cyclists in sample = $\frac{24}{40}$ = ...

Estimate of number of cyclists in year 11 = ... × 250 = ... **(2)**

6 An online company asked people to fill in a questionnaire.

417 people filled in the questionnaire. 48 of these people were not satisfied with the service they received. 5000 people had used the company in the last month.

Work out an estimate for the number of people out of the 5000 who were not satisfied with the service they received.

.... **(2)**

Stratified sampling

Guided **1** Here are the numbers of children in the top three years of a primary school.

Year	4	5	6	Total
Number of children	60	80	60	200

The head teacher wants to take a sample of 40 children stratified by year.
Work out how many children from year 4 should be in the sample.

> Remember that the proportion of year 4 in the sample must be the same as the proportion of year 4 in the school.

Proportion of year 4 in the school $= \dfrac{60}{200} = $ ....

Number of year 4 in the sample = ... × 40 = ... **(2)**

2 Here are the ages of patients who visited a surgery on Monday.

Age group (years)	0–5	6–10	11–20	Over 20
Number of patients	25	35	65	125

(a) Work out the total number of patients. .... **(1)**

The surgery wants to take a sample of 60 patients stratified by age group.

(b) Work out how many patients from the 0–5 age group should be in the sample.

.... **(2)**

3 The table shows information about the working population of a town.

Stratum	Women in part-time work	Women in full-time work	Men in part-time work	Men in full-time work
Number	187	230	256	452

A researcher wants to take a stratified sample of 50 workers from this population.
The researcher says, 'There should be about twice as many men in full-time work in the sample as there are women in full-time work in the sample'.

The researcher is correct.

(a) Give a reason why the researcher is correct.

....

.... **(1)**

(b) Work out how many women in full-time work should be in the sample.

.... **(2)**

Further stratified sampling

Guided 1 A council plans to build a new road next to a village. The new road will make it quicker for people to get to work. It will also lead to an increase in noise and pollution.
The council decides to send a questionnaire to 100 people in the village.

(a) Explain why a stratified sample based on age, with a stratum from 20 years to 55 years and a stratum over 55 years would be suitable.

.... **(1)**

There are 287 people from 20 to 55 years old and 112 people over 55 years old.

(b) How many people in the over 55 years old group should receive a questionnaire?

Proportion of over 55 year olds in the village is $\dfrac{112}{287 + 112}$

so the proportion in the sample should be ....

so the number of over 55 year olds in the sample should be

.... **(2)**

Aiming higher 2 A hospital wants to follow up patients to see if they were satisfied with the treatment they received.

The table gives information about the patients seen at the hospital in the last month.

Age (years) / Gender	Under 30	Between 30 and 50	Over 50
Male	213	412	360
Female	178	384	840

The hospital decides to take a sample of 160 patients stratified by age and gender.
Work out the number of females over 50 who should be in the sample.

.... **(3)**

3 An organisation has this management structure.

Management grade	Senior	Middle	Junior
Number	380	500	620

A stratified sample of 50 managers is to be taken from the organisation.
Work out how many of each management grade there should be in the sample.

.... **(3)**

Further sampling methods

> **Guided**

1 Sue wants to find out what people think of the new speed bumps on a street of 200 houses. She selects the 3rd, 8th, 13th houses and so on.
   (a) Describe this sampling method.

   ....

   (b) Explain **one** advantage and **one** disadvantage of this method.

   Advantage   *Good coverage*

   Disadvantage   *May be affected by patterns*   **(3)**

2 Ali is a marketing assistant for a chocolate company. To find out what people think of a new chocolate bar, he asks 60 women and 20 men.
   (a) Describe this sampling method.

   ....

   (b) Give **one** advantage and **one** disadvantage of this method.

   Advantage...

   Disadvantage.. **(3)**

3 Explain the difference between stratified sampling and cluster sampling.

   > Your answer should refer to randomness and practicality.

   ....

   .... **(2)**

4 Alex is carrying out a survey in the morning by a very busy road. About 80 cars per minute pass the point where Alex is standing.

   He wants to record the licence plate and the number of occupants in 50 cars.

   > Give one practical advantage of systematic sampling that refers to the context of the question.

   (a) Explain why systematic sampling may be easier for Alex than random sampling.

   .... **(1)**

   (b) Work out how long Alex should allow to collect his data.   .... **(1)**

   Alex decides to collect data for half an hour every four hours, starting at 8:30 am and finishing at 5:00 pm.
   (c) Explain why this method of sampling may not be a good one.

   .... **(1)**

5 Anushka is a senior police officer.

   She wants to investigate morale at the 48 police headquarters based all over the UK.
   She knows that about 40 people work at each headquarters.
   She wants to survey about 200 people and ask them what they think of the current situation.
   Anushka takes a random sample of people.
   (a) Give **one** disadvantage of the method she has chosen.

   .... **(1)**

   (b) Describe an alternative sampling method that Anushka could use to improve her survey.

   .... **(2)**

Sampling overview

1   Here is a list of data collection methods.

Census   Random sample   Stratified sample

Cluster sample   Quota sample   Systematic sample

The table shows some data collection situations.
Complete the **Data collection method** column in the table.
Use terms from the list above. They may be used more than once.

Data collection situation	Data collection method
(a) A town council wants to know people's opinions of new parking regulations. The council contacts everyone who lives in the town.	Census
(b) A biologist studies bee populations. He picks out five areas in the country to do a bee census.	
(c) A doctor tries out a new treatment for warts. It is claimed it works better on young people. He asks 5% of young people treated in the last month and 5% of older people treated in the same month whether the treatment worked.	
(d) A town planner wants to find out people's opinions of new lights in a street. He asks every 5th person he meets in the street.	
(e) A marketing manager wants to find out what soap people use. He asks 30 older women and 10 young women.	

(5)

2   The UK National Census takes place every 10 years.

(a) Give **three** reasons why there is a 10-year gap between censuses.

1   A census collects information from the whole population so a lot of information is

collected. This is..

2   ....

3   ....

It is a legal requirement that every householder fills in a census form.
(b) Explain why.

....

In 2011, the census form could be completed online or by a paper form.
(c) Give **two** reasons why the Government wants more people to fill in the form online.

1   ....

2   .... (6)

Data capture sheets

1   Anya wants to collect information about how many male adults and how many female adults eat at her restaurant on Monday night.

Design a suitable data capture sheet for Anya to use.

.... (2)

Guided

2   Jenny made a list of the common spring flowers she saw.

Crocus	Daffodil	Hyacinth	Tulip	Tulip
Hyacinth	Tulip	Daffodil	Crocus	Daffodil
Daffodil	Hyacinth	Tulip	Hyacinth	Tulip
Daffodil	Daffodil	Daffodil	Hyacinth	Tulip
Hyacinth	Tulip	Hyacinth	Hyacinth	Crocus

> Draw 3 columns. The first column should have the names of the plants, one in each row. Put the other two column headings in next.

Record this information in a suitable data capture sheet.

Type of plant		
Crocus		

(3)

3   Mark wants to collect information about the numbers of swimmers and non-swimmers there are in a town and also whether they are adults or children.

Here are the 20 people he asked.

Adult swimmer	Adult non-swimmer	Child non-swimmer	Adult swimmer
Child swimmer	Adult non-swimmer	Child non-swimmer	Adult swimmer
Child swimmer	Child swimmer	Adult non-swimmer	Adult non-swimmer
Adult non-swimmer	Adult swimmer	Adult swimmer	Child swimmer
Child swimmer	Child swimmer	Child swimmer	Child non-swimmer

(a) Record this information in a suitable data capture sheet.

> You will need 1 row for headings and then 4 more rows.

(3)

(b) Which was the largest group? .... (3)

Interviews and questionnaires

1 A council wants to investigate what people do with their rubbish.

One question on their questionnaire is: 'What do you do with your rubbish?'

> A good question in a questionnaire should be unbiased, easy to answer and unambiguous.

(a) Why is this not a good question?

.... **(1)**

Another question on their questionnaire is:
'Don't you think the council does an excellent job in collecting rubbish?'

(b) Why is this not a good question?

.... **(1)**

2 A police community officer wants to find out what young people do in the evenings.
He wants to find out information from 50 young people.
He can choose to interview young people or give out questionnaires.

(a) Give **one** advantage of using interviews.

.... **(1)**

(b) Give **two** disadvantages of using interviews.

1 ....

2 .... **(2)**

He decides to use a questionnaire.
One question is: 'What do you do in the evening?'

(c) Give **one** reason why this is a poor question.

.... **(1)**

The police officer decides to carry out a pilot survey first.

(d) Give **two** reasons for doing a pilot survey.

> Pilot surveys are usually small and are used to test out ideas.

1 ....

2 .... **(2)**

> **Guided** **3** A doctor wants to find out how well people understand the nature and progress of their illness.

(a) Give **one** reason why using an interview might be suitable.

Many illnesses and treatments can be very complex so ....

.... **(1)**

(b) One question that could be asked is: 'How long have you had your illness?'
Explain why this is not a good question.

.... **(1)**

Questionnaires

Guided 1 Valerie is the manager of a supermarket. She wants to find out how often people shop at her supermarket. Design a suitable question for Valerie to use on a questionnaire.

You must include some response boxes.

> A good question should be clear and include a time frame. Response boxes should be clear and cover all possibilities.

How many times do you shop at this supermarket each ....?

☐   ☐   ☐   ☐

Never   Once   ....

(3)

2 Poppy wants to find out how much people use their computer.

She uses this question on a questionnaire.

How much time do you use your computer?

☐   ☐   ☐   ☐   ☐   ☐

0–1 hour 1–2 hours 2–3 hours 3–4 hours 4–5 hours 5–6 hours

> Remember to look at the question and the response boxes.

Write down **two** things that are wrong with this question.

1 ....

2 .... **(2)**

3 James wants to find out how many text messages people send.
He uses this question on a questionnaire.

How many text messages do you send?

☐   ☐   ☐   ☐

1 to 10   11 to 20   21 to 30   More than 30

(a) Write down **two** things that are wrong with this question.

1 ....

2 .... **(2)**

(b) Write an improved question. ....

.... **(2)**

4 Shiree wants to find out how far her football team's supporters travel to get to away matches. Design a suitable question for Shiree.

You must include some response boxes.

(3)

Capture/recapture

1 90 fish were caught from a lake and marked. They were then put back into the lake. One week later 50 fish were caught, of which 12 were found to be marked.

(a) What proportion of the second sample were found to be marked? .... **(1)**

(b) Work out an estimate for the number of fish in the lake.

.... **(1)**

(c) Write down one assumption made in your answer to part (b).

> There are several possible answers. You only have to give one.

That marking a fish does not ....

.... **(1)**

2 A wildlife expert wants to estimate the number of deer in a large forest.
In one week he catches and marks 72 different deer.
Two weeks later he catches 54 deer, of which 7 are found to be marked.
Estimate the number of deer in the forest.

Let N be the number in the population.

> Use this equation, which says that the proportion of marked deer in the second sample is the same as the proportion of marked deer in the population.

$$\frac{N}{72} = \frac{54}{7} \quad \text{so } N = \frac{72 \times 54}{7} = \text{.....................}$$ **(2)**

3 A biologist is studying the population of barnacle geese in the Solway Firth.
In 2012 she catches and rings 320 barnacle geese.
One week later she catches 84 barnacle geese, of which 15 are found to be ringed.

(a) Work out an estimate for the population in 2012.

.... **(2)**

In 2013 she catches 90 barnacle geese, of which 12 are found to be ringed.
(b) Work out the difference in population between 2012 and 2013.

.... **(2)**

(c) Give **one** possible reason why the answers to parts (a) and (b) are different.

.... **(1)**

4 Aled catches, marks and releases 80 fish.
Later he catches 50 fish and finds one of these to be marked.

(a) He wants to estimate the size of the fish population using this sample.
Give **one** reason why this estimate might be unreliable.

.... **(1)**

Aled releases all the fish.
Later he catches 100 fish and finds that 8 are marked.
(b) Work out an estimate for the size of the population.

.... **(3)**

Frequency tables

Guided   1   The table gives information about the ages of people using the village hall one day.

Age in years	0–18	19–30	31–40	41–50	51–60	61–
Frequency	4	12	23	16	8	17

(a) How many people in the sample were younger than 31?

4 + 12 = ...   **(1)**

(b) How many people in the sample were older than 40?

16 + ... + ... + ... = ...   **(1)**

(c) The last column in the table is headed 61–
Explain what this heading means.

....

.... **(1)**

2   The incomplete table gives information about how long it took students to wrap a parcel.

Time, (T) seconds	Tally	Frequency
$0 < T \le 10$	|||	
$10 < T \le 20$	||||	
$20 < T \le 30$	̶|̶|̶|̶|	
$30 < T \le 40$	|||	

> $0 < T \le 10$ means that T can take any value up to and including 10.

Here are 10 further times in seconds: 18   10   12   23   24   30   38   15   9   29

(a) Complete the tally and frequency columns.   **(2)**

(b) Work out the number of students who took more than 20 seconds.

> 'More than 20 seconds' means that you must not include 20 seconds.

.... **(3)**

3   A housing association carried out a survey of the numbers of children in each family in their houses. The results are shown in the table. (No families had more than 5 children.)

Number of children	0	1	2	3	4	5
Frequency	8	12	13	7	3	0

(a) What was the largest number of children in a family?   .... **(1)**

(b) How many families were in the survey?   .... **(1)**

(c) How many children were there altogether?   .... **(2)**

Two-way tables

Guided 1 Some men and women were asked whether they had had a cold in the last two months. The two-way table shows the results.

	Had a cold	Not had a cold	Total
Men	38	43	
Women	20		56
Total			

(a) Complete the table.

Total men = 38 + 43   Total people = ... + 56 **(2)**

(b) How many men were asked altogether?

.... **(1)**

(c) More people had not had a cold than had had a cold. How many more?

> You need to use the totals at the bottom of the table.

.... **(1)**

2 52 students went on a music summer school.
Each student had to choose their favourite session from Drums or Guitar or Vocals.

18 of the students chose Guitar.
21 of the students were girls.
15 of the girls chose Vocals.
None of the girls chose Drums.
5 more boys chose Drums than chose Vocals.

(a) Complete the two-way table.

	Drums	Guitar	Vocals	Total
Girls				
Boys				
Total				52

(3)

(b) How many boys chose Drums? .... **(1)**

3 A teacher asks 60 students if they walk to school, take the bus or travel some other way.

34 of the students are boys.
8 of the boys take the bus.
14 girls walk.
6 of the 10 students travelling some other way are boys.

> This can be done best by using a two-way table.

Work out the number of students who take the bus.

.... **(3)**

Pictograms

Guided   1   The pictogram gives information about the numbers of rugby balls sold in a shop in January, February and March.

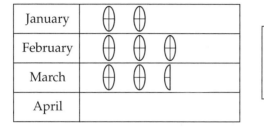 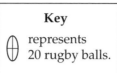

(a) How many rugby balls were sold in January? 20 × 2 = ... **(1)**

(b) How many rugby balls were sold in March? 20 × 2 + 20 ÷ 2 = ... **(1)**

35 rugby balls were sold in April.

(c) Complete the pictogram.   5 = 20 + 10 + 5   **(1)**

2   Helen carried out a survey to find out the number of mobiles that students had in school. Here are her results.

Class	Number of mobiles
Mr Smith	20
Mr Zaheer	10
Ms Lee	16
Mrs Linski	5

Draw a pictogram to show Helen's results.

Use a key of ▯ = 4 mobiles

3   The pictogram gives information about the numbers of laptops sold from a shop in each of 4 months.

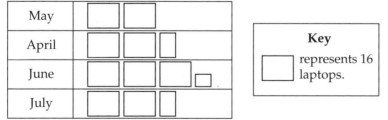

(a) How many laptops were sold in May? .... **(1)**

(b) How many laptops were sold in June? .... **(1)**

The shop made a profit of £125 on each laptop.

(c) Work out the total profit for the 4-month period.

.... **(2)**

Bar charts and vertical line graphs

1   A student did a survey of the favourite colour of each student in his class. He drew a bar chart in a hurry. There are several things wrong with the chart.

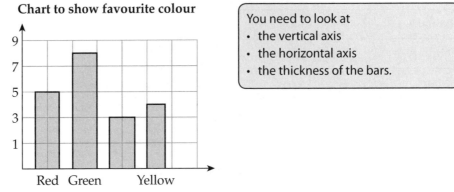

Chart to show favourite colour

You need to look at
• the vertical axis
• the horizontal axis
• the thickness of the bars.

Write down **three** things that are wrong with this bar chart.

1.   ....

2.   ....

3.   .... **(3)**

2   Helen carried out a survey to find out the fruit her friends like best.

Here are her results.

Apple	Orange	Peach	Banana	Pineapple
Banana	Banana	Orange	Apple	Peach
Banana	Orange	Pineapple	Orange	Banana
Peach	Apple	Banana	Apple	Banana

(a) Complete the table for Helen's results.

Fruit	Tally	Frequency
Apple		
Banana		
Orange		
Peach		
Pineapple		

(2)

(b) Draw a suitable chart to show Helen's results on graph paper.   **(2)**

3   The diagram shows a bar chart produced by the owner of company B.

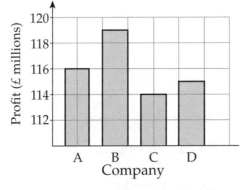

Give **one** reason why the bar chart is misleading.

.... **(1)**

Stem and leaf diagrams

1 Here is a list of the numbers of orchids found in 21 squares, each of area 4 m², in a marsh.

15	8	0	12	21	7	30	23	17	10
4	0	2	14	28	31	22	27	14	17	4

(a) Use these data to complete the stem and leaf diagram below.

```
0 |
1 | 5
2 |
3 |
```

> Start with the stem – the numbers run from 0 to 3.
> Fill in the leaves (including the 0s).
> Include a **key**!

(3)

(b) What fraction of the squares had more than 20 orchids? .... **(1)**

2 A scientist spends two weeks catching moths for research.

The stem and leaf diagram gives information about the numbers he caught on the first ten nights.

```
0 | 6 7
1 | 2 2 3
2 | 0 7 8
3 | 3 4
```

Key: 3 | 4 means 34 moths

On the next four nights he caught 8 moths, 13 moths, 26 moths and 36 moths.

(a) Complete the diagram. **(1)**

(b) Work out the fraction of nights on which he caught more than 12 moths. .... **(1)**

3 James was carrying out an investigation into the lengths of letters published in a newspaper. The list shows the numbers of words in 20 letters written to the newspaper.

126	157	138	165	175	129	138	145	153	128
149	156	171	156	137	132	155	149	170	157

(a) Show this information on a stem and leaf diagram.

(3)

(b) What percentage of the letters had more than 160 words? .... **(2)**

Pie charts

Guided 1 The pie chart shows how a household spent its income in one year.

Household spending in one year

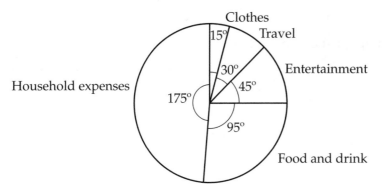

(a) What fraction of the household's spending was on clothes? .... **(1)**

The total household income was £32 000.

(b) Work out how much money was spent on food and drink.

> The fraction of spending on food and drink is $\frac{95}{360}$.
> So you need to find this fraction of the total spending in pounds.

Angle is 95° out of 360°

Amount spent on food and drink is $\frac{95}{360}$ × £32 000 .... **(2)**

2 The pie chart shows information about some people's favourite radio station.

Radio station

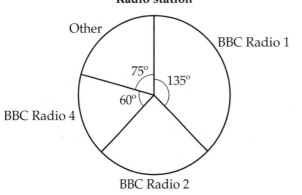

24 people said BBC Radio 2 was their favourite station.

Work out how many people said BBC Radio 4 was their favourite station.

.... **(3)**

Drawing pie charts

Guided 1 The table shows how people took exercise.

Type of exercise	Number of people
Regular walking	240
Cycling	180
Visiting the gym	120
Swimming	100
Other	80

First find the total.
This total is represented by the full circle of 360°.

Draw a fully labelled pie chart to show this information.

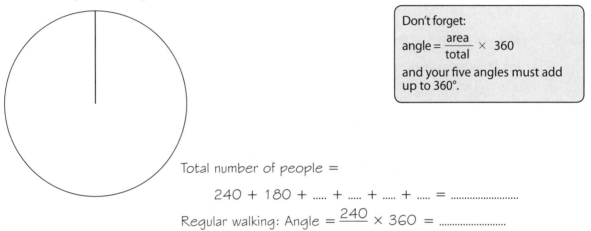

Don't forget:

$$\text{angle} = \frac{\text{area}}{\text{total}} \times 360$$

and your five angles must add up to 360°.

Total number of people =

240 + 180 + ... + ... + ... + ... = ...

Regular walking: Angle = $\frac{240}{\text{..........}}$ × 360 = ...

(3)

2 The table shows the destinations of people in an airport.

Destination	Number of people
Europe	78
America	100
Africa	30
Asia	32

Draw a fully labelled pie chart to show this information.

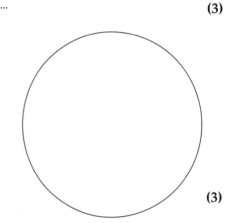

(3)

3 The table gives information about land use for agriculture in Wiltshire.

Type of land use	Area (1000s of hectares)
Cereals	81
Green crops	25
Grass	103
Set aside	37

Draw a fully labelled pie chart to show this information.

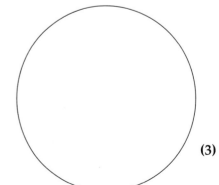

(3)

Bar charts

Guided **1** The multiple bar chart shows the numbers of boys, girls and adults at a sports club each week for four weeks.

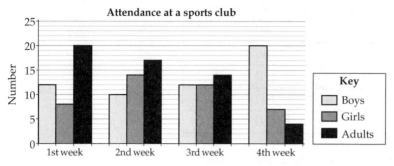

(a) Describe the trend of adult attendance during the four weeks.

The overall pattern of adult attendance from the 1st week to the 4th week is .... **(1)**

(b) In which week were boy and girl attendances the same?

The boys' and the girls' bars are the same height in .... **(1)**

(c) Work out the total attendance in the 4th week.   .... **(1)**

2 This table gives information about employment in the South West for two different years.

	1980	2010
Service	23%	39%
Retail	37%	32%
Other	40%	29%

Draw a composite bar chart to show this information.

(3)

3 The composite bar chart shows sales from a shop over four months.

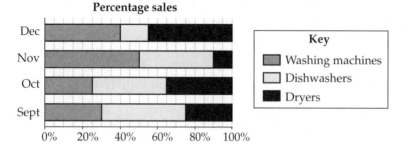

(a) Write down as a percentage the sales of washing machines in September. .... **(1)**

(b) Write down as a percentage the sales of dishwashers in October.   .... **(1)**

Total sales in December were 200.

(c) Work out the number of dryers sold in December.

> Remember to find x% of 200.
> Work out $\frac{x}{100} \times 200$

.... **(2)**

Pie charts with percentages

Guided   **1**   The table gives information about non-skilled employment in the South West for two different years.

	Services	Retail	Other
1980	23%	37%	40%
2010	39%	32%	29%

(a) Draw pie charts to show this information.

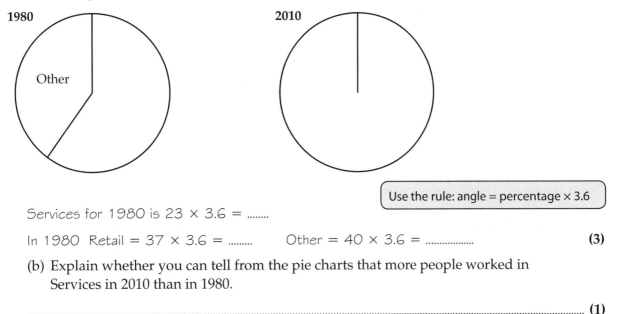

1980   2010

Other

> Use the rule: angle = percentage × 3.6

Services for 1980 is 23 × 3.6 = ...

In 1980 Retail = 37 × 3.6 = ...   Other = 40 × 3.6 = ...   **(3)**

(b) Explain whether you can tell from the pie charts that more people worked in Services in 2010 than in 1980.

.... **(1)**

2   The pie chart shows the degree choices of a group of sixth-form students.

Degree choice

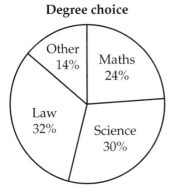

Other 14%   Maths 24%

Law 32%

Science 30%

There were 250 students in the sixth form.
(a) Work out the number of students in each of the four categories.

....

.... **(2)**

Ten of the students who had chosen Law changed their minds and chose Science instead. A new percentage pie chart was drawn to include this change.

> You need to think about the total number of students and the number of students who chose Maths.

(b) Will the angle in the Maths sector of the pie chart change?
     You must give a reason for your answer.

.... **(1)**

Using comparative pie charts

Aiming higher

1 The pie charts show how children in two different primary schools travel to school.

> **Guided**

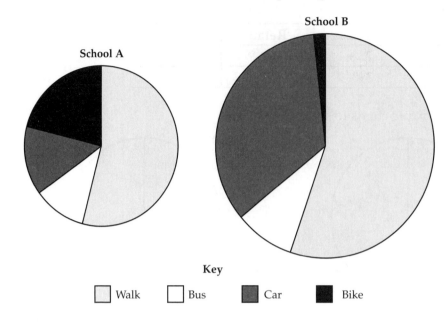

School A

School B

Key

☐ Walk   ☐ Bus   ■ Car   ■ Bike

(a) There are 400 children in school A. How many children are in school B?

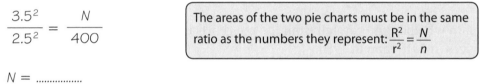

$$\frac{3.5^2}{2.5^2} = \frac{N}{400}$$

The areas of the two pie charts must be in the same ratio as the numbers they represent: $\frac{R^2}{r^2} = \frac{N}{n}$

$N =$ ....

.... **(2)**

(b) How many children from school B walk to school?

.... **(2)**

(c) In which school do more children travel to school by bus?

.... **(2)**

Aiming higher

2 Araminta collected data about use of mobile phones from 40 boys and 90 girls.

She drew a pie chart for the boys with a radius of 5 cm.

Work out the radius she should use for the pie chart for the girls.

.... **(2)**

Frequency polygons

Guided

1 The table gives information about the times that people had to wait in a queue.

Time, t (minutes)	$0 < t \le 1$	$1 < t \le 2$	$2 < t \le 3$	$3 < t \le 4$	$4 < t \le 5$	$5 < t \le 6$
Frequency	7	12	23	17	8	7

(a) Draw a frequency polygon to show this information on graph paper. **(2)**

> Draw and label a horizontal axis from $t = 0$ to $t = 6$.
> Draw and label a vertical axis from 0 to 25.
> Plot each of the points at the midpoint of the interval.
> Join up the points in order with straight lines.

(b) Make one comment about the distribution of times.

The peak of the waiting times occurs .... **(1)**

2 In an experiment to study respiration a group of women were asked to hold their breath for as long as they could after inhaling pure oxygen.

The frequency polygon gives information about the time in seconds the women could hold their breath.

A group of men were also given the same amount of oxygen. The table shows information about how long the men could hold their breath.

Time, t (minutes)	$20 < t \le 25$	$25 < t \le 30$	$30 < t \le 35$	$35 < t \le 40$	$40 < t \le 45$	$45 < t \le 50$	$50 < t \le 55$
Number of men	22	48	64	17	5	3	0

No man could hold his breath for more than 55 seconds.

(a) Plot a frequency polygon for the men's data on the same grid as the women's. **(2)**

> When you compare two frequency polygons you can make comments about the width and where the maximum occurs.

(b) Compare the two distributions.

....

.... **(1)**

23

Cumulative frequency diagrams 1

1 The table gives information about the weekly wages earned by a group of people.

Wage, W (£)	$0 < W \le 200$	$200 < W \le 400$	$400 < W \le 600$	$600 < W \le 800$	$800 < W \le 1000$
Frequency	10	15	18	15	9

(a) Complete the cumulative frequency table for this information.

> You should check that the last entry in the cumulative frequency table is the same as the sum of all the frequencies in the frequency table.

Wage, W (£)	$0 < W \le 200$	$0 < W \le 400$	$0 < W \le 600$	$0 < W \le 800$	$0 < W \le 1000$
Frequency	10	10 + 15 = 25	25 + 18 =		

(1)

(b) Draw a cumulative frequency diagram to show this information.

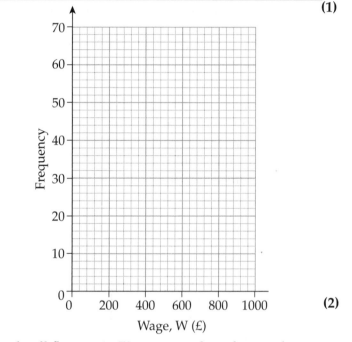

(2)

2 A scientist recorded the numbers of speedwell flowers in 70 squares selected at random in a field. Each square has an area of 1 m².

His results are shown in the frequency table.

Number N per m²	$0 < N \le 10$	$10 < N \le 20$	$20 < N \le 30$	$30 < N \le 40$	$40 < N \le 50$	$50 < N \le 60$
Frequency	13	7	15	15	12	8

(a) Produce a cumulative frequency table for the results.

(1)

(b) Draw a cumulative frequency diagram for the results.

(2)

Cumulative frequency diagrams 2

Guided 1   The table gives information about the numbers of people in a lift on 16 occasions.

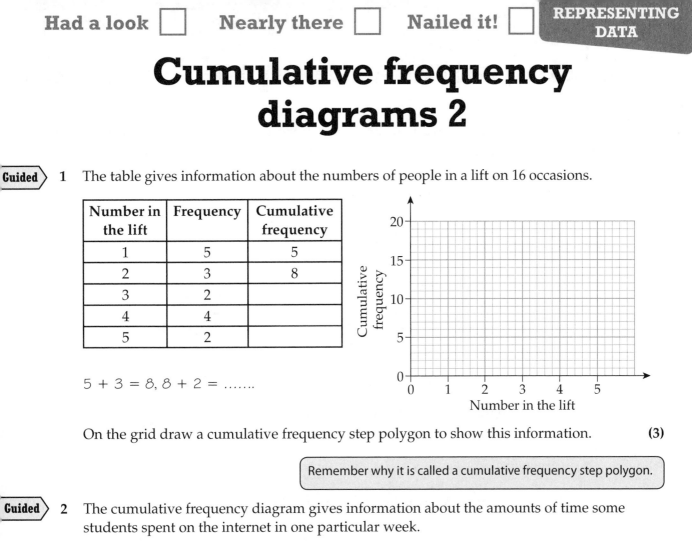

Number in the lift	Frequency	Cumulative frequency
1	5	5
2	3	8
3	2	
4	4	
5	2	

5 + 3 = 8, 8 + 2 = ...

On the grid draw a cumulative frequency step polygon to show this information. **(3)**

> Remember why it is called a cumulative frequency step polygon.

Guided 2   The cumulative frequency diagram gives information about the amounts of time some students spent on the internet in one particular week.

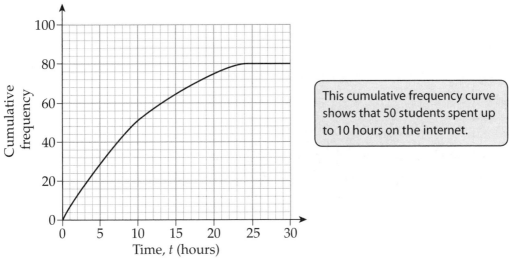

> This cumulative frequency curve shows that 50 students spent up to 10 hours on the internet.

(a) Estimate the number of students who spent at least 15 hours on the internet.

.... **(2)**

(b) Estimate what time on the internet was exceeded by 20 of the students.

The line from 60 on the cumulative frequency axis meets the curve at .... **(1)**

(c) Estimate how many students spent between 10 and 20 hours on the internet.

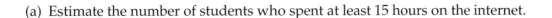

.... **(1)**

Histograms with equal intervals

Guided **1** The table gives some information about the lengths in minutes of some tracks downloaded from an internet site.

Track length, t (minutes)	Number of tracks
$2 \leq t < 3$	4
$3 \leq t < 4$	7
$4 \leq t < 5$	10
$5 \leq t < 6$	8
$6 \leq t < 7$	3

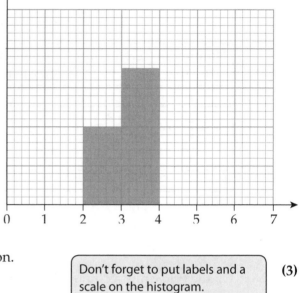

(a) Draw a histogram to show this information.

> Don't forget to put labels and a scale on the histogram.

(3)

(b) What fraction of tracks were longer than 5 minutes? .... **(1)**

2 The histogram gives information about the distribution of heights of children in a swimming club.

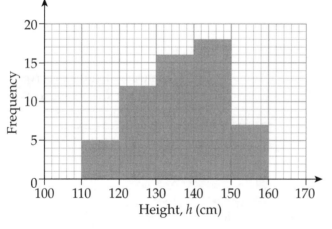

(a) How many children had a height from 120 cm to 130 cm? .... **(1)**

(b) How many children were 140 cm or taller in height?

.... **(1)**

(c) Estimate the number of children with heights between 120 cm and 125 cm.

.... **(1)**

Histograms with unequal intervals

1 Here is some information about the areas of some gardens.

Area, A (m²)	Frequency	Frequency density
$0 < A \le 10$	8	0.8
$10 < A \le 15$	13	$13 \div 5 = \dots$
$15 < A \le 20$	22	$22 \div \dots = \dots$
$20 < A \le 40$	34	
$40 < A \le 100$	18	

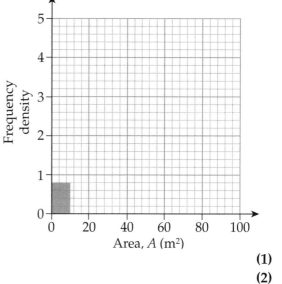

(a) Complete the frequency density column.

(1)

(b) Use the table to complete the histogram.

(2)

2 The table gives information about the costs (£ thousands) of some houses.

Cost, C (£ thousands)	Frequency
$0 < C \le 100$	50
$100 < C \le 200$	130
$200 < C \le 250$	220
$250 < C \le 400$	360
$400 < C \le 1000$	180

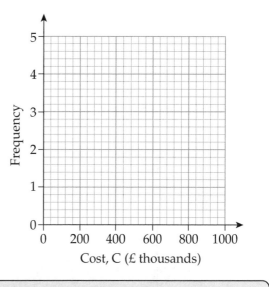

Draw a fully labelled histogram.

Remember: frequency density = frequency ÷ class interval

(3)

3 The table gives information about the ages of patients in a hospital.

Age, A (years)	Frequency
$0 < A \le 10$	50
$10 < A \le 20$	130
$20 < A \le 40$	220
$40 < A \le 60$	360
$60 < A$	180

You have to make a sensible estimate for the upper bound age.

Draw a histogram to show this information.

(3)

Interpreting histograms

 1 This histogram gives information about the times in minutes it took some students to do a maths problem.

Guided

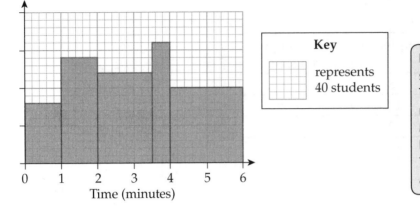

Time (minutes)

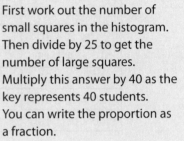

Key

represents 40 students

First work out the number of small squares in the histogram. Then divide by 25 to get the number of large squares. Multiply this answer by 40 as the key represents 40 students. You can write the proportion as a fraction.

Work out the proportion of students who took less than 3 minutes.

$5 \times 8 + 5 \times 14 + 7.5 \times 12 +$ .... small squares

.... $\div 25 \times 40 =$ .... students

.... **(3)**

2 This histogram shows the amount of spending money some students were given each week.

Amount of spending money each week

Amount (£)

20 students were given £4 or less each week.

(a) How many students received between £8 and £12 each week?

.... **(2)**

(b) Estimate the number of students who received between £10 and £20 per week.

.... **(2)**

Population pyramids

Guided 1   The population pyramids of the UK and Iran are shown below.

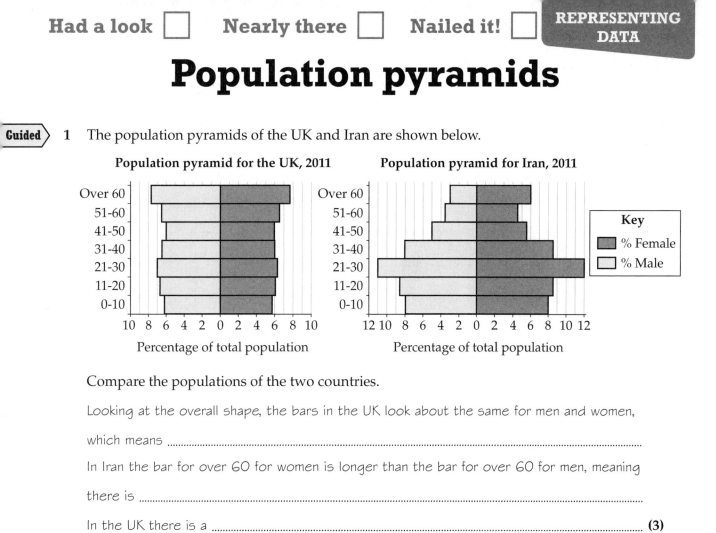

Population pyramid for the UK, 2011   **Population pyramid for Iran, 2011**

Compare the populations of the two countries.

Looking at the overall shape, the bars in the UK look about the same for men and women,

which means ....

In Iran the bar for over 60 for women is longer than the bar for over 60 for men, meaning

there is ....

In the UK there is a .... **(3)**

2   The table gives information about the percentages of males and females by age group in a country.

Males		Females	
Age	Percentage	Age	Percentage
0–20	8	0–20	6
21–40	10	21–40	10
41–60	12	41–60	11
61–80	13	61–80	12
81+	7	81+	11

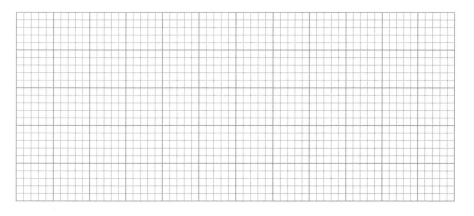

(a) Draw a population pyramid to show this information.   **(3)**

(b) Compare the percentages of males and females in each age group.

.... **(1)**

Choropleth maps

Guided

1 A scientist collected information about the numbers of acorns in some woodland.
The left-hand diagram below shows her results.
(a) Plot a choropleth map of the results. Use the key given below. **(2)**

28	21	11	11	18
15	32	31	29	7
13	36	26	36	16
20	14	6	21	19

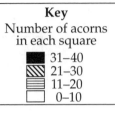

Key
Number of acorns
in each square
■ 31–40
▨ 21–30
▤ 11–20
☐ 0–10

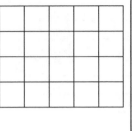

> Start from the top left. Since it's 28, shade it with diagonal lines. Cross out the 28 and move across to the next square and shade with diagonal lines. Cross out the 21 and keep going.

(b) Make a comment about the distribution of acorns.

The distribution of acorns is more ....

.... **(1)**

2 The map shows the counties of South East England.

County	Annual rainfall (mm)
B	638
C	641
E	590
ES	629
H	659
K	610
L	583
NK	622
SK	610
SY	625
WS	634

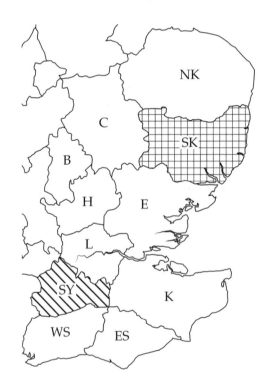

(a) Make a choropleth map of the rainfall by shading in the diagram.

Use the following key:

Key

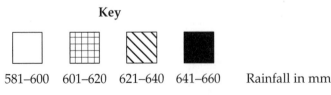

581–600   601–620   621–640   641–660   Rainfall in mm

Two counties have already been done for you. **(2)**

(b) Make a comment about the distribution of rainfall in South East England.

.... **(1)**

Mode, median and mean

Guided 1 Ahmed counted the numbers of cars passing his house at intervals of 1 minute for 15 minutes.

Here are his results: 3   5   8   5   9   0   6   11   7   5   2   7   8   6   5

(a) Work out the mode.

> Write the numbers in order of size. Check you have 15 numbers.

> The MOde is the number of cars which appears MOst.

0 2 3 5...

so .... **(1)**

(b) Work out the median.

The median is the .... th value in the list = ... **(1)**

(c) Work out the mean. **(2)**

Total of all the values = 3 + 5 + ... = ...

Mean = total ÷ .... = ... **(2)**

2 Ten students took a maths test.
Here are their scores.

25   37   22   19   37   43   30   15   34   36

(a) Calculate the mean. .... **(2)**

(b) Calculate the median. .... **(2)**

Later, the score of 15 was changed to 18.
(c) (i) Will the mean increase, decrease or stay the same? Give a reason.

.... **(1)**

(ii) Will the median increase, decrease or stay the same? Give a reason.

.... **(1)**

3 Karl likes motor racing. He gets 5 points for every race he comes first, 3 points for second and 1 point for third. He gets no points for any other placings.
Here are the results of the races he took part in this season.

1st   3rd   4th   6th   2nd   3rd   1st   5th   3rd   4th

Work out the mean number of points that Karl got this season. .... **(2)**

4 Shelagh plays basketball.

These are the points that she has scored so far this season.

> What are the total points she will need, to have a mean of 7 when she has played 9 games?

5 10 8 12 0 7 9 3

How many points will she have to score in her next game to make her mean points per game 7? .... **(2)**

Mean from a frequency table

Guided **1** The table shows information about the numbers of letters in the first 60 words in a children's book.

Work out the mean.

Number of letters	Frequency	fx
1	2	$2 \times 1 = 2$
2	5	$5 \times 2 = 10$
3	14	$14 \times 3 =$
4	19	$\times \quad =$
5	10	
6	10	

> The formula you need is on the formulae sheet.
> Σfx is the total of the fx column.
> Σf is the total of the frequency column.

← Put the total of this column here.

$$\text{Mean} = \frac{\Sigma fx}{\Sigma f} = 240 \div \text{.....} =$$

.... **(3)**

2 The table gives information about the numbers of Christmas presents shoppers had bought one day in a shop.

Number of presents	Frequency	
1	0	
2	8	
3	12	
4	10	
5	6	
6	4	

> Start by using an extra column headed Frequency × number of presents.

Work out the mean number of presents shoppers had bought. .... **(3)**

3 The table gives information about the number of times some people took a driving test until they passed.

Number of times	Frequency
1	18
2	12
3	4
4	3
5	0
6	1

Work out the mean number of times people took a driving test. .... **(3)**

Mean from a grouped frequency table

1 There are 28 people who work in an office.

The table gives information about the distances they travel to work each day.

> Start by adding extra columns headed Midpoint and $f \times$ midpoint.

Distance, d (km)	Frequency	Midpoint	$f \times$ midpoint
$0 < d \le 5$	12		
$5 < d \le 10$	6		
$10 < d \le 15$	4		
$15 < d \le 20$	6		

> Use both empty columns.
> Put the totals at the bottom of the frequency column and the $f \times$ midpoint column.

(a) Work out an estimate for the mean distance an office worker travels to work.

.... km **(4)**

(b) Explain why your answer to part (a) is an estimate.

.... **(2)**

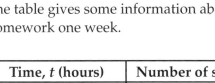

2 The table gives some information about the spending money some students had one week.

Spending money, £S	Number of students	Midpoint	$f \times$ midpoint
$0 < S \le 4$	6	2	$6 \times 2 =$
$4 < S \le 8$	8	8	$8 \times \ldots =$
$8 < S \le 12$	5		
$12 < S \le 16$	3		
$16 < S \le 20$	1		

Work out an estimate for the mean amount of spending money.

£ .... **(4)**

3 The table gives some information about the times a group of students spent on their homework one week.

Time, t (hours)	Number of students
$0 < t \le 2$	5
$2 < t \le 4$	10
$4 < t \le 6$	8
$6 < t \le 8$	5
$8 < t \le 10$	2

Work out an estimate for the mean time spent on homework.   .... **(4)**

Mode and median from a frequency table

> **Guided** **1** The table gives information about the numbers of letters in each word in a newspaper article.

Number of letters	Frequency	Cumulative frequency
1	7	7
2	12	19
3	23	
4	17	
5	13	
6	8	
7	6	

> The mode is the number of letters with the greatest frequency.

(a) Write down the mode. .... **(1)**

(b) Find the median.

$$\frac{n+1}{2} = \frac{...... + 1}{2} = ...$$ The ....th value is ....

.... **(2)**

2 The table gives information about the number of visitors to a small hospital ward for 63 days.

Number of visitors	Frequency
4	4
5	7
6	10
7	17
8	18
9	7

> It's easier to find the median if you work out the cumulative frequencies.

(a) Write down the mode. .... **(1)**

(b) Work out the median. .... **(2)**

On the next visiting day 7 people visited the ward.

(c) (i) What effect will this have on the mode?

....

(ii) What effect will this have on the median?

.... **(2)**

Averages from grouped frequency tables

Guided 1   The table gives information about the lengths of some earthworms found in a field.

Length, L (cm)	Frequency	Cumulative frequency
$0 < L \le 2$	27	
$2 < L \le 4$	22	
$4 < L \le 6$	23	
$6 < L \le 8$	33	
$8 < L \le 10$	18	
$10 < L \le 12$	14	

For the modal class look for the interval with the largest frequency.

(a) Write down the modal class.   .... **(1)**

(b) Work out the class interval that contains the median.

$\dfrac{n + 1}{2} = \dfrac{...... + 1}{2} = ...$ The ......th length is in the interval .... **(2)**

(c) Calculate the mean. Give your answer correct to 3 significant figures.

Median $= .... + \dfrac{\frac{1}{2}(\quad) - ......}{..........} 2 = ...$ .... **(3)**

2   The table gives information about the times that people had to wait in a queue to get served.

Time, T (minutes)	Frequency
$0 < T \le 1$	7
$1 < T \le 2$	12
$2 < T \le 3$	23
$3 < T \le 4$	20
$4 < T \le 5$	14
$5 < T \le 6$	11

(a) Write down the modal class interval.   .... **(1)**

(b) Find the class interval that contains the median.   .... **(2)**

A further 10 people had to wait over 6 minutes in the queue. This group is added to the data in the table.

(c) Which class interval will now contain the median?   .... **(1)**

Although 'over 6 minutes' is not in the table, it is possible to find the position of the new median.

(d) Give a reason for your answer to part (c).

.... **(1)**

Had a look ☐   Nearly there ☐   Nailed it! ☐

Which average?

Guided 1 In a small business of 9 people, the amounts of money they each earn per week are

£225   £215   £200   £210   £200   £225   £210   £230   £420

> First you will need to put the data in order of size, starting with £200.

(a) Find the median. .... **(2)**

(b) Work out the mean. Total ÷ 9 = ... **(2)**

(c) Which average – the mode, median or mean – is the most suitable to use here?
Give reasons for your answer.

> The best way to give your reasons is to say why the other two averages are not suitable.

....

.... **(2)**

2 The table shows the marks of the students in Mrs Smith's class test.

Mark	Number of students
6	4
7	8
8	5
9	3
10	0

> Remember to use cumulative frequency to find the median.

(a) Work out the median mark for the students in the table. .... **(2)**

(b) Calculate the mean mark for the students in the table. .... **(3)**

Mrs Smith had left out the marks of two students new to the class. They had marks of 0 and 1.

(c) What effect would including the marks of these two students have on

(i) the median .... **(1)**

(ii) the mean? .... **(1)**

3 The stem and leaf diagram gives the weights in grams of 15 baskets of fruit.

```
18 | 9 9
19 | 0 3 5 8
20 | 0 5 6 7 7
21 | 1 2 8
22 | 3
```

Key: 18 | 9 represents 189 g

> Don't forget to include the stem when you find the median.

(a) Work out the median weight. .... **(1)**

An extra basket of fruit of weight 190 g is to be included in the stem and leaf diagram.

(b) What effect will this have on the median? .... **(1)**

Estimating the median

> **Guided** 1 The table gives information about the numbers of seconds people could balance on one foot.

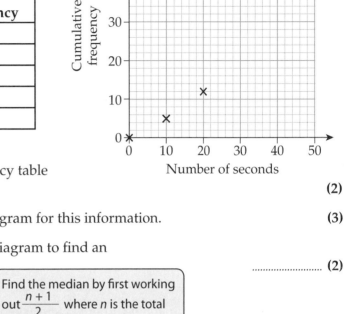

Number of seconds, N	Frequency	Cumulative frequency
$0 < N \leq 10$	5	5
$10 < N \leq 20$	7	12
$20 < N \leq 30$	22	
$30 < N \leq 40$	9	
$40 < N \leq 50$	4	

(a) Complete the cumulative frequency table for these data. **(2)**

(b) Draw a cumulative frequency diagram for this information. **(3)**

(c) Use your cumulative frequency diagram to find an estimate for the median. .... **(2)**

> Find the median by first working out $\dfrac{n+1}{2}$ where n is the total frequency.

2 The table gives information taken from a survey of 45 families about spending on food one week.

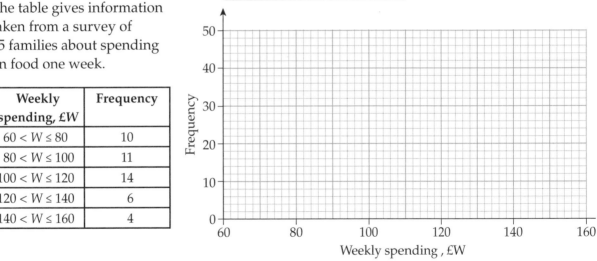

Weekly spending, £W	Frequency
$60 < W \leq 80$	10
$80 < W \leq 100$	11
$100 < W \leq 120$	14
$120 < W \leq 140$	6
$140 < W \leq 160$	4

(a) Draw a cumulative frequency diagram for these data. **(3)**

(b) Use the cumulative frequency diagram to find an estimate for the median. .... **(1)**

Five families had been left out of the survey.
They each spent more than £160 that week.

(c) (i) What effect would the inclusion of these families have on the median?

....

    (ii) Give a reason for your answer.

.... **(2)**

The mean of combined samples

Guided **1** The mean height of a group of 18 male adults is 174 cm.
The mean height of a group of 12 female adults is 162 cm.

Work out the mean height of the people in the combined group.

> Work out the total of the heights of all the males and the females in the combined group.

The total height of the 18 males is 18 × 174 = ...cm

The total height of the 12 females is ....cm

The total height of the 30 people in the combined group is ....

The mean height of the 30 people is .... **(3)**

2 There are 16 boys and 12 girls in a class.

The mean number of text messages sent by the boys yesterday was 5.5.
The mean number of text messages sent by the girls yesterday was 7.

Work out the mean number of text messages sent by the class yesterday.

.... **(1)**

3 A group of 50 university students were asked about the number of cups of coffee
they had drunk the previous day.

The mean number of cups drunk was 3.6.
The mean for the 30 male students was 4.5.
Work out the mean for the 20 female students.

> Subtract the total number of cups drunk by the male students from the total number of cups drunk by all the students.

.... **(3)**

4 In 2012 the mean weekly rainfall in an area was 69.4 mm.

The mean weekly rainfall in the same area in the first 20 weeks of 2013 was 65 mm.
What would the mean weekly rainfall have to be in the remaining
32 weeks of 2013 to give the same mean as in 2012?

.... **(3)**

5 Aptitude tests are taken by people when they apply for some jobs.

For the applicants in group A the mean mark on an aptitude test was 80%.
For the applicants in group B the mean mark on the aptitude test was 45%.

40% of the applicants who took the test came from group A.

Work out the mean percentage mark for the whole group of applicants.

.... **(3)**

Weighted means

Guided ➤ 1 In an exam, the mark on Paper 1 has a weighting of 40% and the mark on Paper 2 has a weighting of 60%.

Any student getting a mean mark of 65 or more will pass the exam.

Kate gets 75 on Paper 1 and 55 on Paper 2.

(a) Will Kate pass the test?
You must give a reason for your answer.

> Use the formula for weighted means:
> $\dfrac{\Sigma wx}{\Sigma w}$ where the ws are the weights and the xs are the marks.
> You will have to learn this formula.

$$\frac{40 \times 75 + 60 \times 55}{40 + 60} = \ldots\ldots\ldots\ldots$$

.... **(1)**

Kumar gets 68 on Paper 1.

(b) What is the lowest mark he can get on Paper 2 and still pass the exam?

.... **(3)**

2 In an entrance test for a company, candidates have to do two tests – a critical thinking test and a numerical test.

The weight given to the critical thinking test is twice the weight given to the numerical test. Here are the scores of a group of 8 candidates.

Candidate	A	B	C	D	E	F	G	H
Critical thinking score	36	58	39	25	48	44	23	12
Numerical score	27	45	21	25	29	43	15	12

> Work out the mean critical thinking score and the mean numerical score first.

The weighted score of candidate A is 33.

Work out the mean weighted score for the group of 8 candidates.

.... **(3)**

3 In a vegetable-growing competition, marks are given out of 10 for size, for shape, for colour and for texture. The marks are then weighted as shown in the table.

Quality	Size	Shape	Colour	Texture
Weight	1	3	2	2

Mr Smith's leeks get 7 marks for size, 8 marks for shape, 6 marks for colour and 10 marks for texture.

(a) Work out the weighted score for Mr Smith's leeks.

.... **(2)**

Mrs Jones's carrots had a weighted score of 7.5.
They got 10 marks for size, 6 marks for shape and 9 marks for texture.

(b) Work out what mark they were given for colour.

.... **(2)**

Measures of spread

1 Work out the range of this set of values.

   1 6 8 7 4 5 7 9 6 4   .... **(1)**

> **Guided**

2 Here is a list of the minimum temperatures in a garden for the first 7 days in January.

   –6°C 2°C 3°C –1°C 0°C 3°C 2°C

   (a) Work out the range of these temperatures.

   > Start by putting the numbers in order. Put the –6°C first.

   Range = largest – smallest = ...

   (1)

   The range of the minimum temperatures in the garden for the first 8 days in January is 11°C.

   (b) What are the **two** possible values for the minimum temperature in the garden on the 8th day in January?

   There are two possible answers: –6 + 11 = ... or .... – .... = ...

   .... or .... **(2)**

3 15 students did a Geography test.

   Here are the marks:

   Boys   19 13 19 12 13   9   9 17
   Girls   20 14 13 15 12 10 13

   James thinks that the range of the boys' marks is 1 more than the range of the girls' marks. Is James correct? You must give a reason.

   .... **(2)**

4 A scientist was studying the effect of caterpillars on the growth of rose bushes.
   He counted the numbers of caterpillars on 11 rose bushes.

   8   1   5   9   2   0   7   1   3   0   2

   (a) Work out the lower quartile.   .... **(2)**

   (b) Work out the upper quartile.   .... **(1)**

   (c) Find the interquartile range.   .... **(1)**

5 This table gives information about the numbers of TVs in people's houses.

Number of TVs	Frequency	
0	5	
1	8	
2	11	
3	8	
4	8	
5	4	
6	3	
7 or more	3	

> You need to complete a cumulative frequency column.

   Find the interquartile range.   .... **(3)**

Box plots

Guided 1 The number of bags of crisps sold each day in a shop was recorded over a 15-day period. The results are shown below.

20   15   10   30   33   40   5   11   13   20   25   42   31   17   26

(a) Find the median and the quartiles of these data.

5 10 11 13 15 17 20 ....

> Start by putting the values in order. Put the smallest value first.

The median is ....

Q_1 = 4th value = ...   and Q_3 = ...   **(3)**

(b) Draw a box plot to show these data.

> Start with the lowest and highest values from the list. Then make the box from Q_1, the median and Q_3.

0   10   20   30   40   50
Number sold   **(2)**

2 The number of antiques a dealer bought each day over a 15-day period is shown in the list.

18   12   13   24   38   23   7   10   23   28   17   13   23   31   15

(a) Find the median and the quartiles.

.... **(3)**

(b) Draw a box plot for these data.

0   10   20   30   40   50
Number bought   **(2)**

3 The stem and leaf diagram gives information about the weights of potatoes dug from some potato plants.

```
0 | 8  8
1 | 2  5  6  7  8
2 | 4  4  7  9
3 | 3  3
```

Key: 2|4 means 2.4 kg

Draw a box plot to show this information.

(3)

41

Interquartile range and continuous data

> **Guided**

1 The table gives information about household incomes in the UK in 2012.

Weekly household income, I (£)	Percentage	Cumulative frequency
$0 < I \leqslant 150$	6	6
$150 < I \leqslant 300$	25	6 + 25 = 31
$300 < I \leqslant 450$	27	31 +
$450 < I \leqslant 600$	19	
$600 < I \leqslant 750$	10	
$750 < I \leqslant 900$	7	
$900 < I \leqslant 1050$	3	
$1050 < I \leqslant 1200$	1	
$1200 < I \leqslant 1350$	2	

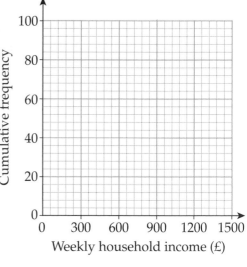

(a) Complete the cumulative frequency table for these data. **(1)**

(b) Draw a cumulative frequency diagram. **(2)**

(c) Find an estimate for the interquartile range. .... **(2)**

2 The cumulative frequency diagram gives information about how much the people in one town spent on entertainment in one weekend.

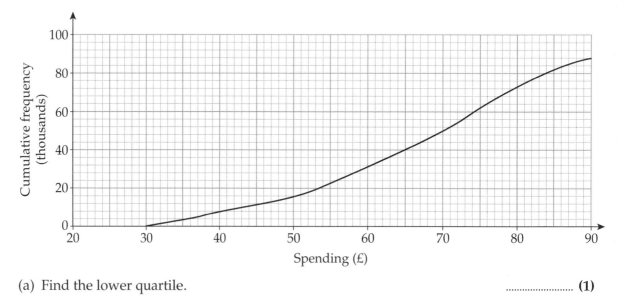

(a) Find the lower quartile. .... **(1)**

(b) Work out an estimate for the interquartile range. .... **(1)**

> $Q_1 = \dfrac{n+1}{4}$ th value but you can use $\dfrac{n}{4}$ as n is so large.

Percentiles and deciles

1   The cumulative frequency diagram gives information about the distribution of UK house prices in 2010 and in 2013.

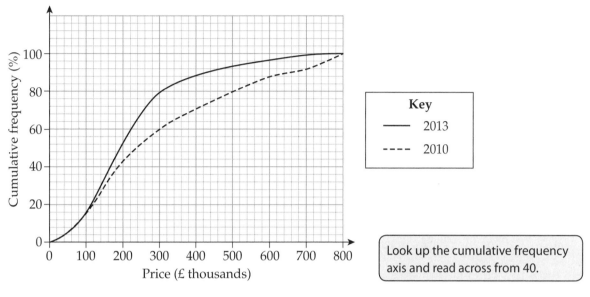

Key
— 2013
---- 2010

> Look up the cumulative frequency axis and read across from 40.

(a) Work out the value of the 40th percentile in 2013.   .... **(1)**

(b) Work out the difference in price between a house at the 8th decile in 2013 and a house at the 8th decile in 2010.

.... **(2)**

Guided ⟩   2   The cumulative frequency diagram gives information about the times runners took to complete the London Marathon in 2012.

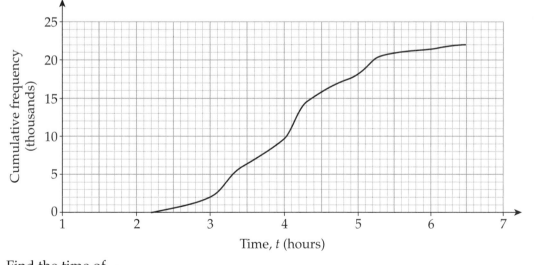

Find the time of
(a) the lower quartile   .... **(1)**

(b) the 45th percentile $\dfrac{45}{100} \times 22000 =$ .... **(1)**

(c) the 7th decile $\dfrac{7}{10} \times 22000 =$ .... **(1)**

(d) Work out the 20th – 80th interpercentile range.

80th percentile – 20th percentile = ... **(1)**

Comparing discrete distributions

Guided 1   A group of men and a group of women were asked how many times they had used their car in the last week.

Here are the results:

Men   2   4   5   5   6   7   8   10   12   12

Women   3   5   12   4   0   2   6   3   8   5   11

(a) Work out the median for the men.   .... **(1)**

(b) Work out the range for the men.

.... **(1)**

(c) Compare the two distributions.

> Use the answers to parts (a) and (b) to help.
> Don't forget to give your answers in context.

The women's list in order is 0 2 3 3 ....

Women's median = ... so, on average, ...

Women's range = ... so .... **(3)**

2   The stem and leaf diagram gives some information about the numbers of television programmes some men watched one week.

```
0 | 3   5   8
1 | 0   2   2   5   7   7
2 | 4   4   5   9
3 | 0   3
```

A group of women also noted the number of television programmes they watched that week. For this group, the median number was 23 and the interquartile range was 8. Compare the two distributions.

....

.... **(2)**

3   The box plot gives information about the number of times boys were late for school. The data for girls over the same period of time is shown in the table.

Minimum value	0
Lower quartile	2
Median	6
Upper quartile	12
Maximum value	15

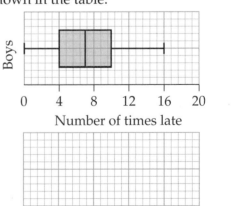

(a) Draw a box plot for the girls.   **(2)**

(b) Compare the distribution of the number of times the boys were late with the distribution of the number of times the girls were late.

....

....

.... **(4)**

Cumulative frequency diagrams and box plots

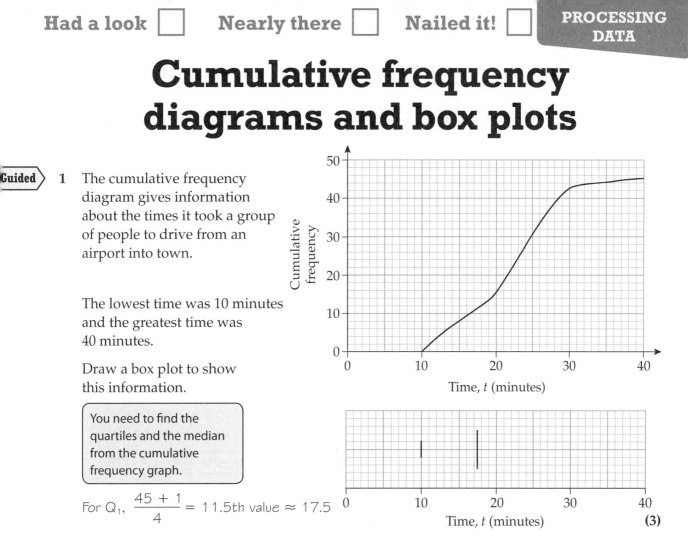

Guided

1 The cumulative frequency diagram gives information about the times it took a group of people to drive from an airport into town.

The lowest time was 10 minutes and the greatest time was 40 minutes.

Draw a box plot to show this information.

> You need to find the quartiles and the median from the cumulative frequency graph.

For Q_1, $\dfrac{45 + 1}{4} = 11.5$th value ≈ 17.5

(3)

2 The cumulative frequency table gives information about the lengths of some earthworms.

Length (cm)	Up to 2	Up to 4	Up to 6	Up to 8	Up to 10	Up to 12
Cumulative frequency	27	49	72	105	123	137

(a) Draw a cumulative frequency diagram to show this information.

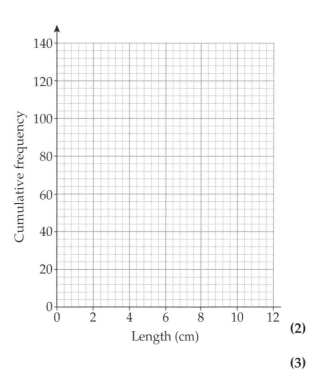

The shortest length was 1.2 cm.
The greatest length was 11.5 cm.

(b) Draw a box plot for the information.

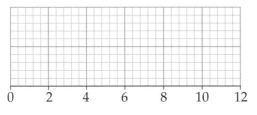

(2)

(3)

Using cumulative frequency diagrams and box plots

Guided 1 The cumulative frequency diagram gives information about the times it took some men to get to work.

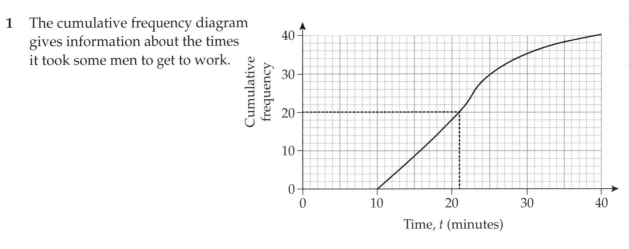

Time, t (minutes)

The lowest time was 10 minutes and the greatest time was 40 minutes.
Similar information about the times for a group of women is shown in the table.

Lowest time	Lower quartile	Median	Upper quartile	Greatest time
8 minutes	13 minutes	26 minutes	31 minutes	45 minutes

Compare the distribution of the times for the women with the distribution of the times for the men.

The median time for the men was .... so on average their times ....

....

The interquartile range ....

.... **(3)**

2 The box plot gives information about the numbers of phone calls some boys made in one week.

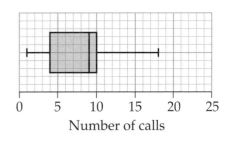

0   5   10   15   20   25
Number of calls

For the girls, the lowest number of calls was 4
and the greatest number was 22.
25% of the girls made no more than 7 calls in the week.
The interquartile range for the girls was 10.
50% of the girls made no more than 12 calls in the week.

(a) Draw a box plot to show this information. **(2)**

(b) Compare the distribution of the calls made by the girls with the distribution of the calls made by boys.

....

....

.... **(2)**

Box plots and outliers

Guided 1 Here is a list of the numbers of different tracks 15 people had downloaded one day.

0 0 2 5 5 6 6 7 7 8 11 11 13 15 21

(a) Determine whether 21 is an outlier.

$Q_1 = $ .... $Q_3 = $ ....

$Q_3 - Q_1 = $ .... – .... = ...

$1.5 \times$ .... = ..., $Q_3 + $ .... = ...

So 21 is ....

> First work out Q_1, Q_2 and Q_3.
> Then calculate $Q_3 + 1.5 \times IQR$
> and $Q_1 - 0.5 \times IQR$

(2)

(b) Draw a box plot to show this information.   **(3)**

Number of tracks

2 The box plots give information about the numbers of text messages sent by some boys and sent by some girls.

(a) Write down the median for

the girls. ....   **(1)**

(b) Write down the interquartile range

for the boys. ....   **(1)**

(c) Show by calculation that there is one outlier for the boys.

Number of texts

....

....

....   **(2)**

3 Alf grows tomato plants. He gave fertiliser to plants in one group (A).
To another group (B) of the same type of plants he gave no fertiliser.
The lists show the number of tomatoes he got from each plant in the two groups.

Group A   5   12   13   13   14   14   17   18   20   22   25

Group B   11   12   12   12   14   15   15   16   16   17   18   19   20

(a) For each group, determine whether there are any outliers.

....

....   **(2)**

(b) Is there any evidence that the fertiliser improved growth?
You must give reasons for your answer.

> If there are any outliers then
> don't use the range to compare.
> Use the median and the IQR.

....

....   **(2)**

Box plots and skewness

Guided **1** Here is a list of the numbers of emails some people received one day.

4   5   6   7   1   5   7   8   11   15   16   11   13   4   6

(a) Draw a box plot to show this information.

Numbers of emails   **(3)**

(b) Does the box plot give evidence for a skew distribution?

$Q_3 - Q_2$ = ... , $Q_2 - Q_1$ = ...

> Look at the position of Q_2 with respect to Q_1 and Q_3 on the box plot.

.... **(1)**

2 The box plots give information about the numbers of times in a term some boys and girls were late for school.

(a) Write down the median for the girls.

.... **(1)**

(b) Write down the interquartile range

for the boys. .... **(1)**

(c) Describe the skew of each of the distributions.

Times late

....

.... **(2)**

3 Johnny got some onions from a shop and some onions from the market.

He weighed the onions.

Here are the masses, in grams, of the onions from the market:

160   120   200   150   195   135   140   120   150   120   110   160   170

The box plot shows information about the masses of the onions from the shop.

(a) Draw a box plot for the masses of onions from the market.

(b) Find the interquartile range for the masses of the onions from the shop.

(3)

Mass, m (grams)

.... **(2)**

(c) Write down **three** comparisons of the distributions. One should be a comparison of the skew.

> The other two comparisons should be of the median and the IQR. Don't forget to interpret your numbers in context.

1. ....

2. ....

3. .... **(3)**

Variance and standard deviation

Aiming higher

Guided

1 Here is a list of the numbers of strawberries obtained from 8 plants in a garden:

> The variance $= \dfrac{\Sigma x^2}{n} - \text{mean}^2$
> where the xs are the numbers of strawberries.

7 9 10 12 8 12 6 0

(a) Work out the variance for these data.

The mean $= 64 \div 8 = \ldots\ldots\ldots\ldots$

Sum of squares $= 7^2 + 9^2 + 10^2 + 12^2 + 8^2 + \ldots\ldots\ldots\ldots\ldots\ldots = \ldots\ldots\ldots\ldots$

Variance $= \ldots\ldots\ldots\ldots\ldots\ldots\ldots\ldots\ldots\ldots$ **(2)**

(b) Work out the standard deviation for these data.

Standard deviation $= \sqrt{\ldots\ldots\ldots\ldots} = \ldots\ldots\ldots\ldots$ **(1)**

Aiming higher

2 The numbers of passengers, x, joining 10 trains in a station were recorded.

You are given that $\sum x = 284$ and $\sum x^2 = 9160$.

> Σx stands for the total number of passengers.
> Σx^2 stands for the sum of the squares of the xs.

(a) Work out the mean number of passengers joining the trains. $\ldots\ldots\ldots\ldots$ **(1)**

(b) Work out the standard deviation for these data.
Give your answer correct to 3 significant figures. $\ldots\ldots\ldots\ldots$ **(2)**

Aiming higher

3 The number of birds, x, in a garden was recorded every 5 minutes for 1 hour.

0 4 6 12 1 0 3 5 5 10 10 7

(a) Work out $\sum x$. $\ldots\ldots\ldots\ldots$ **(1)**

(b) Work out $\sum x^2$. $\ldots\ldots\ldots\ldots$ **(1)**

(c) Work out the standard deviation of the number of birds. $\ldots\ldots\ldots\ldots$ **(2)**

Aiming higher

4 The number of caterpillars on each of 8 cabbage plants was recorded.
The mean was 4.5 and the standard deviation was 2.3.

(a) Work out the variance of the number of caterpillars. $\ldots\ldots\ldots\ldots$ **(1)**

> Remember that variance $= \dfrac{\Sigma x^2}{n} - \text{mean}^2$
> so $\Sigma x^2 = (\text{variance} + \text{mean}^2) \times n$

(b) Work out the value of $\sum x^2$. $\ldots\ldots\ldots\ldots$ **(2)**

Aiming higher

5 Umar measured the widths, w mm, of a sample of 10 leaves.
He found that $\sum w = 500$ and $\sum w^2 = 30000$.

(a) Work out the mean width of the 10 leaves. $\ldots\ldots\ldots\ldots$ **(1)**

(b) Work out the standard deviation of the 10 leaves. $\ldots\ldots\ldots\ldots$ **(1)**

Umar then included two more leaves in his sample. Each leaf had a width of 30 cm.

(c) Work out the mean for all 12 leaves. $\ldots\ldots\ldots\ldots$ **(1)**

(d) Work out the standard deviation for all 12 leaves. $\ldots\ldots\ldots\ldots$ **(1)**

Standard deviation from frequency tables

1 The table gives some information about the sales of a company's cars.

Guided

Sales, S (£000s)	Frequency	Mid S	fS	fS^2
$8 < S \leqslant 14$	80	11	880	9 680
$14 < S \leqslant 20$	120	17		34 680
$20 < S \leqslant 26$	100	23	2300	
$26 < S \leqslant 32$	50	29	1450	42 050

(Source: Prices based on Ford Motor Company)

> You can work out the variance first using
> $$\frac{\Sigma fs^2}{\Sigma f} - \left(\frac{\Sigma fs}{\Sigma f}\right)^2$$

(a) Complete the table.

$120 \times 17 = \text{.........}$ and $100 \times \text{.........} = \text{.................}$ **(1)**

(b) Work out an estimate for the mean.   mean $= \dfrac{\Sigma fs}{\Sigma f} = \text{.................}$ **(1)**

(c) Work out an estimate for the standard deviation.

$\Sigma fs^2 = \text{.................}$

standard deviation $= \sqrt{\dfrac{\Sigma fs^2}{\Sigma f} - \text{mean}^2} = \text{.................}$ **(2)**

2 A scientist was collecting data about the weights of a type of fish.

The partially completed table gives some information about her results.

Weight, W (kg)	Frequency	fW	fW^2
$0 < W \leqslant 1$	10	5	2.5
$1 < W \leqslant 2$	18		
$2 < W \leqslant 3$	14	35	87.5
$3 < W \leqslant 4$	8	28	98
$4 < W \leqslant 5$	2		40.5

> Work out the first missing row from 18×1.5 and 18×1.5^2
> Don't forget, you only square the 1.5.

(a) Complete the table.   **(1)**

(b) Calculate an estimate for the mean.   ....   **(1)**

(c) Calculate an estimate for the standard deviation.

....   **(2)**

3 The partially completed table gives some information about the heights of some children.

Height, h (cm)	Frequency	fh	fh^2
$100 < h \leqslant 110$	8		88 200
$110 < h \leqslant 120$	20	2300	264 500
$120 < h \leqslant 130$	16	2000	250 000
$130 < h \leqslant 140$	5	675	
$140 < h \leqslant 150$	1	145	21 025

Calculate an estimate for the standard deviation.

....   **(4)**

Simple index numbers

Guided   1   The table shows the national minimum wages for workers in the UK.

Year	Adult rate (for workers aged 21+)	Young workers' rate (for 16–17 year olds')
2010	£5.93	£3.64
2004	£4.85	£3.00

(a) Using 2004 as the base year, work out the index numbers for

(i) adult workers in 2010

$$\frac{5.93}{4.85} \times 100 = \text{.........}$$

(2)

(ii) 16–17 year olds in 2010.

(2)

$$\frac{\text{.........}}{\text{.......}} \times 100 = \text{.........}$$

> You need to comment on the two index numbers and on which one, if any, of the two groups has had a greater increase.

(b) Make a comment on your answers. **(1)**

The index number for the adults .... than the index number for the 16–17 year

olds, showing that the adults had a proportionally .... in their minimum wage.

2   The total value of the cars sold from a dealership was £1.8 million in 2011.

Using 2011 as the base year, the index number for the total value of the cars sold in 2012 was 91.

(a) Work out the total value of the cars sold in 2012.   £ .... **(2)**

The total value of the cars sold in 2013 was £1.17 million.

(b) Using 2011 as the base year, work out the index number for 2013.   .... **(2)**

3   The Retail Price Index (RPI) is one way of measuring price inflation.

The base year is 1987, and in 2012 the RPI was 231.5.
A company has always paid wages to its workers in line with the RPI, so if the RPI doubled then so would the wages.
A worker at the company was paid £220 per week in 1987.

> Remember: The base year always has an index number of 100.

(a) Work out what the worker was paid in 2012.   £ .... **(2)**

A manager at the company was paid £328 per week in 1987 and £720 per week in 2012.
The manager thought that he was being paid too little in 2012.

(b) Was the manager correct? You must give a reason for your answer.

....

.... **(2)**

Chain base index numbers

Guided   1   The table shows the value of a house each year for four years.

Year	2008	2009	2010	2011
Price (£ thousands)	105	108	113	117
Chain base index number	100			

(a) Complete the table. The base year is 2008.

> You calculate a chain base index number by using this year's and the previous year's values.

2009 $\dfrac{108}{105} \times \text{........} = \text{....................}$

2010 $\dfrac{113}{\text{........}} \times 100 = \text{....................}$

2011 $\dfrac{\text{........}}{\text{........}} \times \text{........} = \text{....................}$ **(4)**

(b) Which year had the greatest percentage increase in value compared with the previous year?
Give a reason for your answer.

....

.... **(1)**

2   The table shows the average earnings of people in work over three years.

Year	2004	2005	2006
Earnings (£)	32 333	35 018	37 704
Chain base index number	100		

> Chain base index numbers allow you to compare percentage increases year by year.

(a) Taking 2004 as the base year, work out the chain base index numbers for 2005 and 2006.

.... **(2)**

(b) Explain why these figures may be deceptive in deciding how well earnings are rising.

....

.... **(1)**

> Don't forget that chain base index numbers compare year-on-year values.

Weighted index numbers

Guided   **1**   A manufacturer makes animal feed from brewer's yeast and wheat germ.
The ratio is 1 tonne of brewer's yeast to 2 tonnes of wheat germ.
The table shows the prices in 2010 and in 2013.

Year	2010	2013
Price (£) per tonne of brewer's yeast	1250	1600
Price (£) per tonne of wheat germ	1100	1320

> Work out each separate index number first.
>
> Then use $\dfrac{\Sigma wi}{\Sigma w}$ where the *w*s are the weights and the *i*s are the index numbers.

(a) Using 2010 as the base year, work out the weighted index number for 2013 for the animal feed.

Brewer's yeast index number $= \dfrac{1600}{1250} \times 100 =$ ....

Wheat germ index number $= \dfrac{\text{........}}{\text{........}} \times 100 =$ ....

The weights are 1 for brewer's yeast and .... for wheat germ.

The weighted index is $\dfrac{1 \times \text{.......} + \text{.......}}{1 + 2} =$ ....   **(3)**

In 2010 the Consumer Prices Index (CPI) was 115 and in 2013 it was 125.

(b) Compare the change in the weighted index with the change in the CPI.

.... **(2)**

2   Alice has shares in three companies, A, B and C.

The percentages of the numbers of shares she has are shown in the first table.

Company	A	B	C
Percentage of shares	50	30	20

The values of the shares for two years are shown in the second table.

Company	A	B	C
Value of one share in 2010	£2.00	£3.50	£1.78
Value of one share in 2013	£2.18	£2.98	98p

> You have to work out the three simple index numbers first. Then use the percentages as the weights.

Work out the weighted index number for the total value of Alice's shares in 2013.
Use 2010 as the base year.

.... **(3)**

Standardised scores

Aiming higher

1 Flora took a numerical test and a verbal reasoning test.

On the numerical test she got a mark of 48.
On the verbal reasoning test she got a mark of 56.

Here is some information about the two tests.

> You need to learn the formula:
>
> $$\text{Standardised score} = \frac{\text{mark} - \text{mean}}{\text{standard deviation}}$$

Test	Mean	Standard deviation
Numerical	43	5
Verbal reasoning	58	8

(a) Work out Flora's standardised scores on
   (i) the numerical test

Standardised score for the numerical test $= \dfrac{48 - 43}{5} = $ ....

   (ii) the verbal reasoning test.

Standardised score for the verbal reasoning test $= \dfrac{56 - \text{.......}}{\text{.......}} = $ .... **(6)**

Vanessa also took the tests.
Her standardised scores were 2.4 for numerical and −1.5 for verbal reasoning.

> You can do this with some easy algebra.
> Let N be Vanessa's mark for the numerical test.
> Then $\dfrac{N - 43}{5} = 2.4$

(b) Work out Vanessa's marks on
   (i) the numerical test .... **(1)**

   (ii) the verbal reasoning test. .... **(1)**

(c) Compare the performance of Flora and Vanessa on the tests.

....

.... **(2)**

Aiming higher

2 80 people took an abstract reasoning test and a symbol sorting test.
Information about the marks they got is given in the table.

Abstract reasoning mark (x)	$\sum x = 4160$	$\sum x^2 = 234\,320$
Symbol sorting mark (y)	$\sum y = 5200$	$\sum y^2 = 346\,000$

> Work out the mean and the standard deviation for each test first. Then work out the standardised scores.

Lee got a mark of 60 in the abstract reasoning test and a mark of 68 in the symbol sorting test.
Compare his performance on the two tests.

....

....

....

.... **(4)**

Scatter diagrams and correlation

Guided **1** Use the best words from the list to describe each of the correlations shown.

Strong	Weak	Negative	Positive

(a)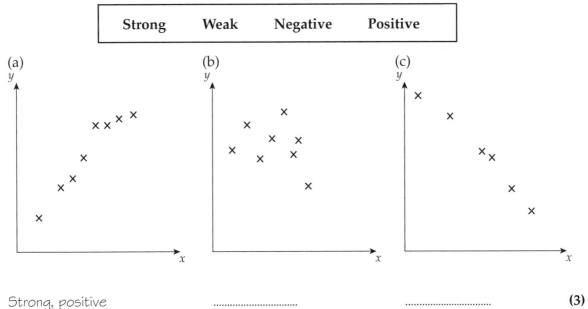

(b)

(c)

Strong, positive   .... **(3)**

2 The table shows the relationship between the engine size and the CO_2 emissions of 8 cars.

Engine size (litres)	1	1.2	1.6	1.6	1.9	2.1	2.3	2.3
Emissions (g per km travelled)	138	143	165	158	185	199	220	211

(a) Which is the explanatory (independent) variable?   .... **(1)**

> Look at the first pair of values to help you decide.

(b) Plot a scatter diagram of the data with the explanatory variable on the horizontal axis. **(2)**

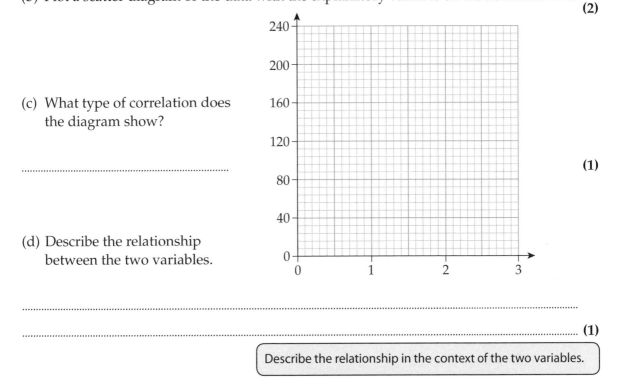

(c) What type of correlation does the diagram show?

.... **(1)**

(d) Describe the relationship between the two variables.

....
.... **(1)**

> Describe the relationship in the context of the two variables.

Lines of best fit

1   A scientist is studying the amount of liquid lost from puddles through evaporation.

He makes 10 puddles of the same size and depth.

He keeps the puddles at different temperatures and measures the amount of liquid lost over a week.

Temperature (°C)	3	5	6	8	10	12	13	15	16	18
Liquid lost (ml)	36	50	53	61	69	79	82	90	88	96

(a) Draw a scatter diagram to represent these data. **(3)**

> Don't forget that the independent (explanatory) variable goes on the horizontal axis.

(b) Work out the mean temperature and the mean amount of liquid lost.

.... **(2)**

(c) Draw a line of best fit on your diagram. **(1)**

(d) Estimate the amount of liquid lost at a temperature of 9°C.

.... **(1)**

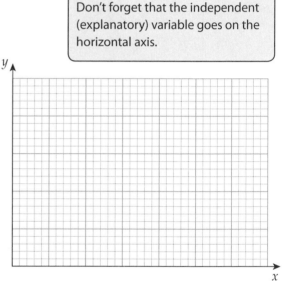

> **Guided**   2   The scatter diagram gives information about the height (x cm), and the mass (y kg), of 10 students.

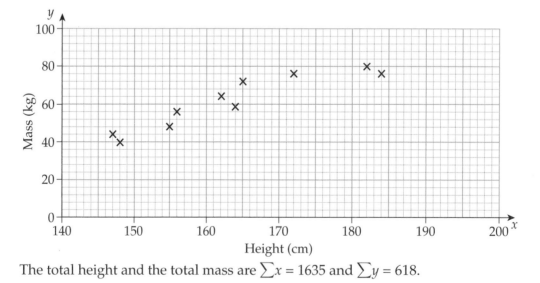

The total height and the total mass are $\sum x = 1635$ and $\sum y = 618$.

(a) Work out the mean value of x and the mean value of y.

$\bar{x} = 6135 \div 10 = $ .... **(2)**

(b) Draw a line of best fit on the diagram. **(1)**

(c) Use your line of best fit to estimate the mass of a student of height 150 cm.

.... **(1)**

Interpolation and extrapolation

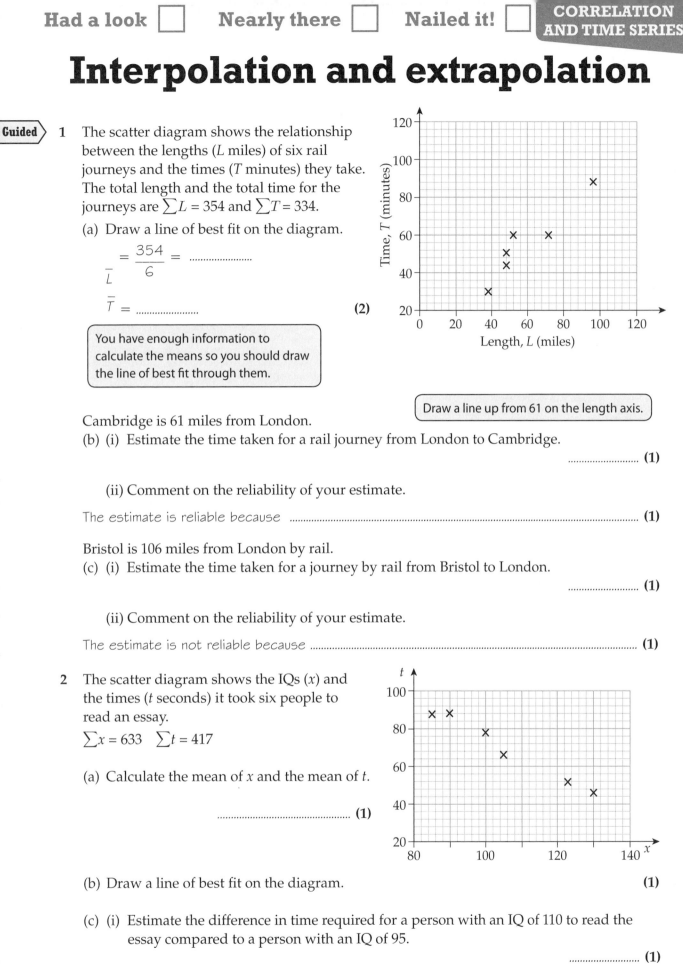

Guided 1 The scatter diagram shows the relationship between the lengths (L miles) of six rail journeys and the times (T minutes) they take. The total length and the total time for the journeys are $\sum L = 354$ and $\sum T = 334$.

(a) Draw a line of best fit on the diagram.

$$\overline{L} = \frac{354}{6} = \text{.....................}$$

$$\overline{T} = \text{.....................} \quad \textbf{(2)}$$

> You have enough information to calculate the means so you should draw the line of best fit through them.

> Draw a line up from 61 on the length axis.

Cambridge is 61 miles from London.

(b) (i) Estimate the time taken for a rail journey from London to Cambridge.

.... **(1)**

(ii) Comment on the reliability of your estimate.

The estimate is reliable because .... **(1)**

Bristol is 106 miles from London by rail.

(c) (i) Estimate the time taken for a journey by rail from Bristol to London.

.... **(1)**

(ii) Comment on the reliability of your estimate.

The estimate is not reliable because .... **(1)**

2 The scatter diagram shows the IQs (x) and the times (t seconds) it took six people to read an essay.

$$\sum x = 633 \quad \sum t = 417$$

(a) Calculate the mean of x and the mean of t.

.... **(1)**

(b) Draw a line of best fit on the diagram. **(1)**

(c) (i) Estimate the difference in time required for a person with an IQ of 110 to read the essay compared to a person with an IQ of 95.

.... **(1)**

(ii) Make a comment about the reliability of your estimate.

....

.... **(3)**

The equation of the line of best fit

1 The scatter diagram gives information about the costs of some rail journeys and their durations.

(a) Find the equation of the line of best fit. Give your answer in the form $y = ax + b$

> a is the gradient and b is the intercept on the y-axis. To work out the gradient, draw a triangle and work out
> $$\text{gradient} = \frac{\text{height}}{\text{base}}$$

$a = \dfrac{\dots\dots\dots}{\dots\dots\dots} = \dots\dots\dots$

$b = 15$

(3)

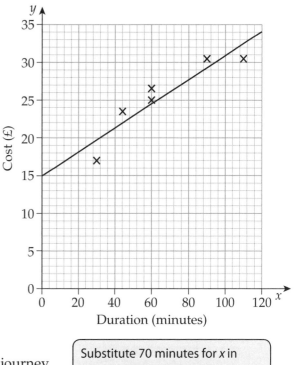

(b) Use your equation to estimate the cost of a journey with a duration of 1 hour 10 minutes.

> Substitute 70 minutes for x in your equation to find the cost, £y.

1 hour 10 minutes = 70 minutes

£ .... **(2)**

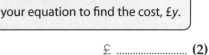

2 The scatter diagram gives information about the mass of some books and the number of pages they have. A line of best fit has been drawn on the diagram.

(a) Find the equation of the line of best fit.

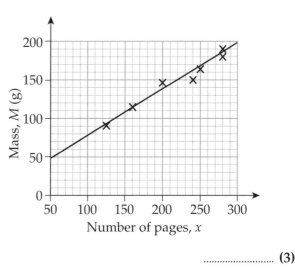

.... **(3)**

> The intercept cannot be read off the graph so use the equation in the rearranged form of $b = y - ax$ to work out the value of b.

(b) Use your equation to estimate the mass of a book with 180 pages. .... **(1)**

Curves of best fit

Aiming higher

1 A student is investigating the speed of flow of water from a tank as it empties. She measures the speed (v) for various initial depths (d) of water in the tank.

The scatter diagram shows her results.

(a) Show that the curve with equation
$$v = 0.44\sqrt{d}$$
gives a good fit to the points.

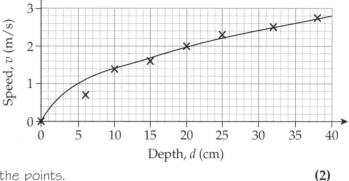

d	0	10	20	30	40
v	0	1.4			

The curve .... shows a good fit to the points. **(2)**

To check the curve, she carries out two further measurements.

> Draw the full curve on the graph to check the fit with the points rather then checking the individual points.

Depth, d (cm)	28	50
Speed, v (m/s)	2.4	2.8

(b) Comment on whether these further measurements support her equation or not.

$0.44\sqrt{28} =$ .... $0.44\sqrt{50} =$ ....

The further measurements .... **(3)**

Aiming higher

2 A scientist is studying how the size of a population of cells changes over time.

The diagram gives information about the size of the population, P, at various times, x minutes.

(a) Show that the curve with equation
$$P = 50 \times 2^x$$
gives a good fit to the points.

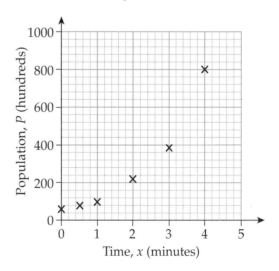

> You will need to make a table of values.
> Use values of x from 0 to 5. Plot the points and draw a smooth curve through them.

(2)

(b) (i) Use the equation of the curve to estimate the time when the population would first exceed 1000.

.... **(1)**

(ii) Comment on the reliability of your answer to part (i).

.... **(1)**

Spearman's rank correlation coefficient 1

1   In a poetry competition, two judges ranked the entries, with 1 as their favourite and 6 as their least favourite.

Entry	A	B	C	D	E	F
Judge A	1	2	3	4	5	6
Judge B	2	4	1	5	6	3

(a) Work out Spearman's rank correlation coefficient for these data.
    You may use the blank row in the table to help with your calculations.   .... **(3)**

> First find the differences in the ranks.

(b) Interpret your answer to part (a).

.... **(1)**

2   In an experiment on problem-solving skills, a group of students were given a verbal task and a numerical task. The order in which they finished each task is given in the table.

Student	A	B	C	D	E	F	G
Verbal task	1st	2nd	3rd	4th	5th	6th	7th
Numerical test	4th	3rd	1st	2nd	6th	7th	5th

(a) Work out Spearman's rank correlation coefficient.
    You may use the blank row in the table to help with your calculations.   .... **(3)**

The designer of the experiment thought that there would be a strong relationship between the students' abilities on the two tasks.

> You need to think what the value from part (a) means in this context, and whether it supports the original expectation.

(b) Is this supported by the results?
    Give a reason for your answer.

.... **(1)**

Aiming higher

3   Alice and Jeff were asked to rank some perfumes, based on how well they liked them. Their rankings are given in the table.

Perfume	A	B	C	D	E	F	G	H
Alice	1	2	3	4	5	6	7	8
Jeff	4	3	6	2	1	7	5	8

(a) Work out Spearman's rank correlation coefficient.
    You may use the blank row in the table to help with your calculations.   .... **(3)**

Afterwards, they discovered that they had used the rankings differently. Alice had given 1 to her favourite, and Jeff had given 8 to his favourite.

(b) What effect would this have on the value calculated in part (a)?

.... **(1)**

Spearman's rank correlation coefficient 2

Aiming higher

Guided

1 A scientist measured the time it took a group of people to complete some jigsaw puzzles. She plotted the time on the scatter diagram against the number of pieces in the puzzle.

The table also gives information about the numbers of pieces and the time it took to complete the jigsaws.

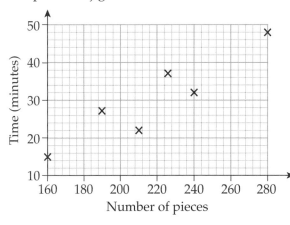

Number of pieces	Time to complete (minutes)		
160	15		
190	27		
210	22		
225	37		
240	32		
280	48		

Use the two columns to enter the ranks of each of the variables.

(a) Work out Spearman's rank correlation coefficient for these data.

You may use the blank columns in the table to help with your calculations. .... **(3)**

(b) Does the value you found in part (a) agree with the pattern shown in the scatter diagram?

The Spearman's rank correlation coefficient found is ....

.... **(1)**

Aiming higher

2 A scientist conducted a study of the number of hours of sunshine and the number of bees seen in a garden for 7 days. The table shows the results of the study.

Number of hours of sunshine	Number of bees	Rank for sunshine	Rank for bees
8	6		
2	0		
0	1		
4	3		
5	5		
7	10		
10	13		

(a) Calculate Spearman's rank correlation coefficient. .... **(3)**

(b) Explain how your answer to part (a) supports the hypothesis that more bees come out on sunny days.

.... **(1)**

The requirements for the study change so that the sunshine is measured in minutes instead of hours.

e.g. 8 hours becomes 480 minutes.

(c) What difference, if any, would this make to the answer found in part (a)?

.... **(1)**

Time series

1 A nurse recorded the temperature of a patient every two hours. This is shown on the time series graph.

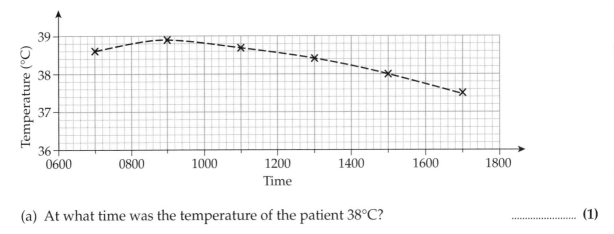

(a) At what time was the temperature of the patient 38°C?   .... **(1)**

(b) What was the highest temperature recorded?   .... **(1)**

(c) During which two-hour period did the temperature of the patient increase?

.... **(1)**

> **Guided**

2 The table gives information about the number of hours of sunshine in a town each month for six months.

Month	Jan	Feb	Mar	Apr	May	Jun
Hours of sunshine	56	62	65	76	84	91

(a) Plot a time series graph to show this information. **(2)**

(b) Describe how the number of hours of sunshine changes throughout the six months.

The hours of sunshine start at 56 in January and then they ....

.... **(2)**

> Describe the overall pattern in context, making sure you give suitable values.

Moving averages

1 The table gives information about the number of house sales made by an estate agent during each six-month period for four years.

Year	Season	Number of sales	2-point moving averages
2010	Summer	20	
	Winter	42	
2011	Summer	18	
	Winter	40	
2012	Summer	18	
	Winter	40	
2013	Summer	18	
	Winter	40	

(a) Work out the 2-point moving averages.

(2)

(b) Comment on any trend in the sales.

.... (1)

2 The table gives information about the total rainfall every four months in a region. Some of the 3-point moving averages have been already been calculated.

Year	2011			2012			2013		
Rainfall (cm)	102	156	142	106	157	135	110	168	143
3-point moving average		133	135	135	133	134			

(a) Complete the table to show the last **two** 3-point moving averages. (2)

> Make a comment in context about the moving averages.

(b) Comment on the trend in the total rainfall.

....

.... (1)

Moving averages and trend lines

Guided 1 The table shows information about the catch of fish, in thousands, from a Pacific port and the 5-point moving averages.

Year	2003	2004	2005	2006	2007	2008	2009	2010	2011	2012	2013
Catch (000s)	29	24	28	32	30	23	25	29	33	31	25
5-point moving average			28.6	28.5	28.25	27.5	26.75	27.5	29.5		

The data for the catches each year have been plotted on the time series graph below.

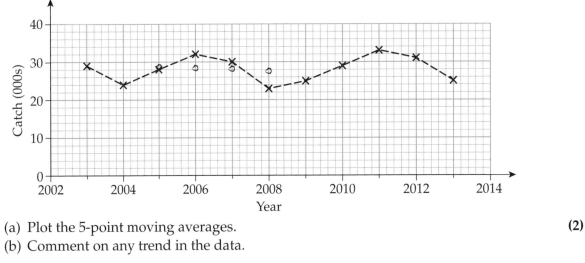

(a) Plot the 5-point moving averages. **(2)**

(b) Comment on any trend in the data.

The moving averages show that .... **(1)**

2 An ice cream seller kept a record of his ice cream sales each quarter for the years 2010, 2011 and 2012.

Year	2010				2011				2012			
Quarter	1	2	3	4	1	2	3	4	1	2	3	4
Sales (000s)	23	31	38	28	25	32	39	28	26	33	41	30
			30	30.5	30.75	31	31	31.25	31.5	32	32.5	

(a) Plot the sales on the time series graph. **(2)**

(b) Plot the moving averages. **(2)**

> The first moving average is plotted between quarter 2 and quarter 3.

(c) Draw a trend line on the graph. **(1)**

> Draw a line of best fit for the moving averages.

(d) Describe the trend of ice cream sales.

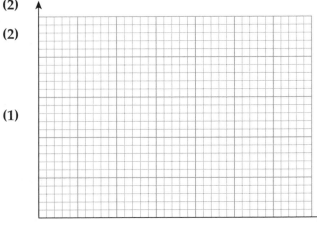

.... **(1)**

The equation of the trend line

1 The diagram shows the production, in thousands, of bananas on a plantation. It also shows the 4-point moving averages and the trend line.

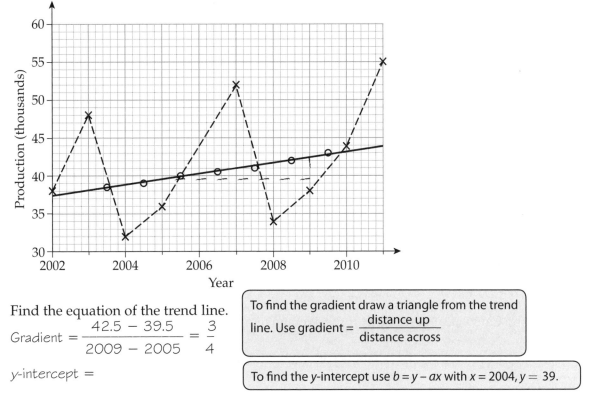

Find the equation of the trend line.

Gradient $= \dfrac{42.5 - 39.5}{2009 - 2005} = \dfrac{3}{4}$

y-intercept $=$

> To find the gradient draw a triangle from the trend line. Use gradient $= \dfrac{\text{distance up}}{\text{distance across}}$

> To find the y-intercept use $b = y - ax$ with $x = 2004$, $y = 39$.

Equation of the line is .... **(2)**

2 The diagram shows the variation of mean temperature by quarter in the UK from 2011 to 2013.

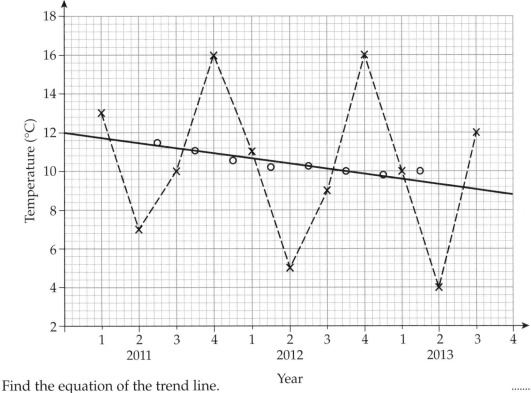

Find the equation of the trend line.   Year   .... **(3)**

Had a look ☐   Nearly there ☐   Nailed it! ☐

Prediction

Aiming higher

Guided

1 The time series graph shows the average temperature by season in Oslo, Norway, over three years. It also shows the 4-point moving averages.

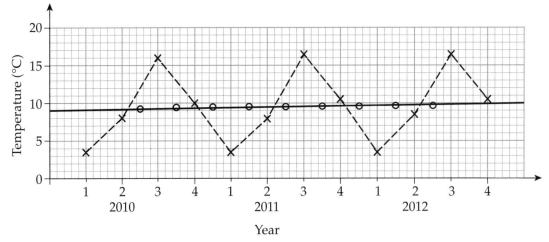

(a) Work out the mean seasonal variation for the 3rd quarter.

$(7 + 7 + \text{......}) \div 3 = \text{..............................}$ **(1)**

(b) Calculate an estimate for the average temperature in the 3rd quarter of 2013.

.... **(3)**

(c) Comment on the reliability of your estimate.

....

> Use your calculation from part (a) to see how much above the trend line the 3rd quarter temperature is.

(1)

Aiming higher

2 The time series graph shows the variation in sales (in thousands) by a company over several years (W = winter, S = summer). It also shows the trend line.

> Read off the graph to find the variation in any season. For the mean summer variation there are four summer variations to find.

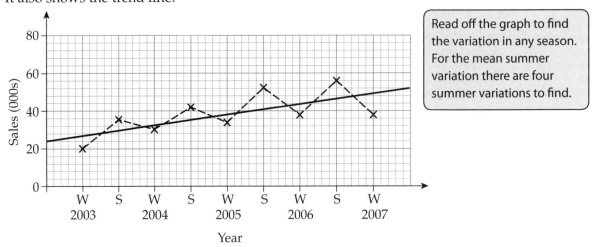

(a) Calculate the mean variation for summer sales. .... **(2)**

(b) Use your answer to part (a) to predict the summer sales in 2007. .... **(2)**

(c) Predict the winter sales for 2008. .... **(2)**

(d) Which of the predictions in parts (b) and (c) is the more reliable? Give a reason for your answer.

.... **(1)**

Probability

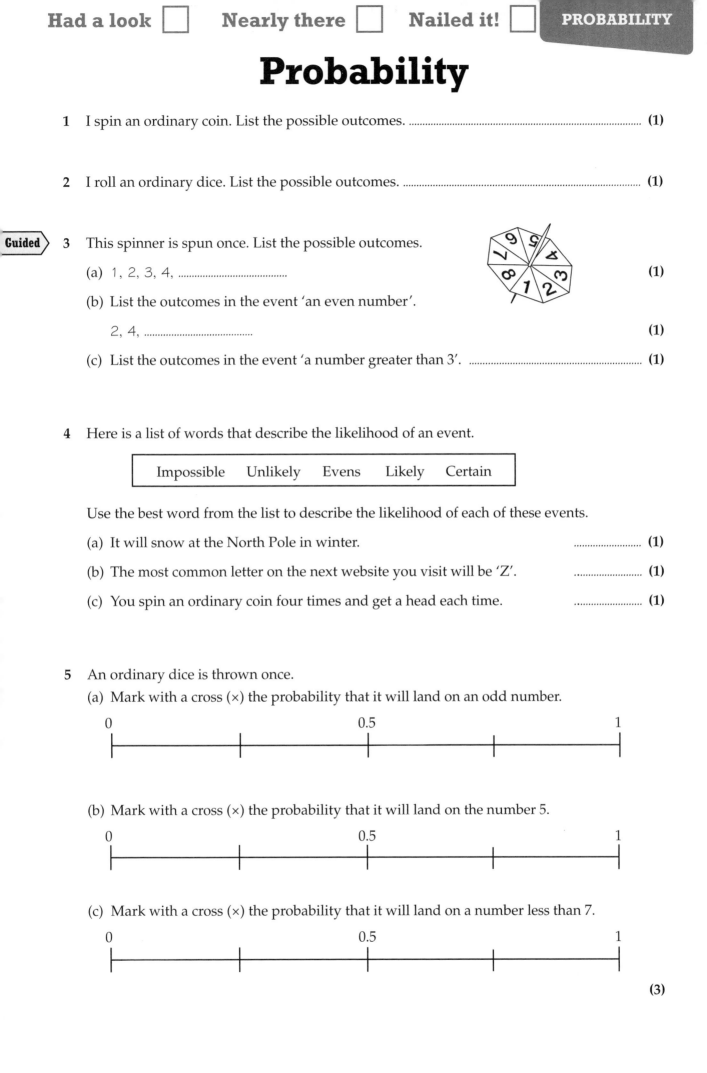

1 I spin an ordinary coin. List the possible outcomes. .... **(1)**

2 I roll an ordinary dice. List the possible outcomes. .... **(1)**

Guided **3** This spinner is spun once. List the possible outcomes.

(a) 1, 2, 3, 4, ... **(1)**

(b) List the outcomes in the event 'an even number'.

     2, 4, ... **(1)**

(c) List the outcomes in the event 'a number greater than 3'. .... **(1)**

4 Here is a list of words that describe the likelihood of an event.

> Impossible   Unlikely   Evens   Likely   Certain

Use the best word from the list to describe the likelihood of each of these events.

(a) It will snow at the North Pole in winter. .... **(1)**

(b) The most common letter on the next website you visit will be 'Z'. .... **(1)**

(c) You spin an ordinary coin four times and get a head each time. .... **(1)**

5 An ordinary dice is thrown once.
(a) Mark with a cross (×) the probability that it will land on an odd number.

0   0.5   1

(b) Mark with a cross (×) the probability that it will land on the number 5.

0   0.5   1

(c) Mark with a cross (×) the probability that it will land on a number less than 7.

0   0.5   1

(3)

Sample spaces

1   A fair coin is spun once. Write down all the possible outcomes. .... **(1)**

2   A fair coin is spun twice. Write down all the possible outcomes. .... **(1)**

⟩ **Guided** ⟩ 3   A fair coin and a fair dice are each thrown once.

(a) Write down all the possible outcomes.

(H, 1) .... **(2)**

(b) How many outcomes have both a head and a 2? .... **(1)**

4   Three fair coins are spun. One possible outcome is HHH (all three coins land on heads).

(a) Write down the other seven possible outcomes. .... **(1)**

Harry says that when he spins three fair coins, he is more likely to get three heads than two heads and a tail.

> Look at the number of outcomes which give three heads and the number of outcomes which give two heads and one tail.

(b) Is Harry correct? You must give a reason for your answer.

.... **(1)**

⟩ **Guided** ⟩ 5   Here are two fair spinners.
Each spinner is spun once.

(a) List all the pairs that make up the sample space.

(1, 1), (1, 2), (1, 3) .... **(1)**

Spinner A   **Spinner B**

(b) How many pairs in the sample space have a total of 3? .... **(1)**

> One pair is (1, 2).

(c) Which total is the most likely? .... **(1)**

> Look for the total that occurs most often.

6   Here are two unusual but fair spinners.
Each spinner is spun once.

(a) List all the pairs that make up the sample space.

.... **(1)**

Spinner X   **Spinner Y**

(b) How many pairs in the sample space have a total of 3? .... **(1)**

(c) Which total is the least likely? .... **(1)**

Probability and sample spaces

1   There are 3 green beads, 2 white beads and 4 red beads in a bag.

A bead is selected at random.

(a) Write down the sample space.

.... **(1)**

(b) Work out the probability that the bead will be | You do not have to cancel any fractions. |

     (i) green .... **(1)**   (ii) red .... **(1)**   (iii) white or red. .... **(1)**

2   Two fair spinners are each numbered 1 to 4.

Each spinner is spun once.

(a) Write down the sample space.

.... **(1)**

(b) Write down the probability that the spinners land on the same number. .... **(1)**

(c) Find the probability that the sum of the numbers is 6. .... **(1)**

3   (a) Complete this sample space to show all the possible outcomes when two ordinary dice are each rolled once.

     (1, 1)   (1, 2)   (1, 3)   (1, 4)   (1, 5)   (1, 6)

     (2, 1)   (2, 2)   (2, 3)   (2, 4)   (2, 5)   (2, 6)

     (3, 1)   (3, 2)   (3, 3)   (3, 4)   (3, 5)   (3, 6)

     (4, 1)   (4, 2)   (4, 3)   (4, 4)

                                                                (1)

(b) Write down the probability that the dice both land on 6. .... **(1)**

(c) Work out the probability that the difference between the numbers the dice land on is 1.

                                                                    .... **(2)**

4   Here are three cards.

     2   **3**   **4** | One possible pair is (2, 2). |

A card is picked at random and put back.

Again, a card is picked at random and put back.

(a) Write down all the possible sequences of two numbers in the sample space.

.... **(2)**

(b) Work out the probability that the same card is selected both times. .... **(1)**

(c) Work out the probability that the sum of the numbers picked is 6. .... **(1)**

Venn diagrams and probability

 1 A group of people were asked whether they get their news from TV or the internet. The Venn diagram gives information about their replies.

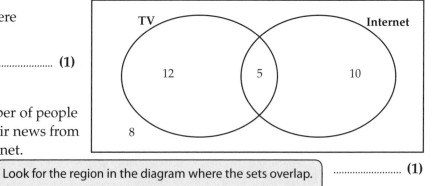

(a) How many people were asked altogether?

12 + 5 + 10 + ... = ... **(1)**

(b) Write down the number of people who said they get their news from both TV and the internet.

| Look for the region in the diagram where the sets overlap. |

.... **(1)**

(c) How many people said they get their news from the internet only? .... **(1)**

Aiming higher **2** A group of 30 students were asked whether they liked football and whether they liked rugby.

20 said they liked football.
12 said they liked football AND rugby.
17 said they liked rugby.

> You will need to draw two overlapping ovals. The '12' goes in the overlap. There were 17 – 12 students who liked rugby but not football.

(a) Draw a Venn diagram to show this information. **(2)**

(b) Work out the number who said they liked neither game. .... **(1)**

(c) Work out the probability that a student selected at random liked football but not rugby.

.... **(1)**

Aiming higher **3** A group of 36 students were asked what pets they own.

15 own a cat.
16 own a dog.
7 own a rabbit.
7 own a cat and a dog.
2 own a cat and a rabbit.
3 own a dog and a rabbit.
1 student owns all three types of animal.

> Draw three overlapping ovals. Put the number 1 in the overlap of all three ovals. Then work your way outwards.

(a) Draw a Venn diagram to show this information.

(4)

(b) Find the probability that a student selected at random owns a cat but not a dog or a rabbit. .... **(1)**

Mutually exclusive events

1 Ben has a biased dice.
When he rolls the dice once, the probability that he gets a 6 is 0.2.
What is the probability that he does not get a 6? .... **(1)**

2 Shola has a biased coin.
When he spins the coin once, the probability that he gets a head is 0.4.
What is the probability that he does not get a head? .... **(1)**

3 Kirstie has a biased spinner.

She spins the spinner once.
The table gives information about the probability of
the spinner landing on each number.

Number	1	2	3	4
Probability	0.3	0.25	0.2	x

> Use the fact that the sum of the probabilities in the table must be 1.

(a) Work out the value of x.

$x = 1 -$ .... $=$ .... **(1)**

(b) Work out the probability that the spinner
lands on the number 1 or the number 3.

> These two outcomes are mutually exclusive.

.... **(1)**

4 There are only red beads, white beads and green beads in a bag.

The probability of picking a red bead is 0.3, and the probability of picking a white bead
is 0.25. Karl picks a bead at random from the bag.

(a) Work out the probability that the bead is

  (i) either red or white .... **(1)**

  (ii) not white .... **(1)**

  (iii) either white or green. .... **(1)**

Karl puts the bead back so that there are 20 beads in the bag.
He adds more white beads so that the number of white beads in the bag is doubled.
Karl takes a bead at random from the bag.

> Work out the total number of beads in the bag first.

(b) Work out the probability that the bead is

  (i) either red or white .... **(2)**

  (ii) not white .... **(2)**

  (iii) either white or green. .... **(2)**

Independent events

1   Naomi has a fair spinner with three sides numbered 1, 2 and 3.

> As the spinner is fair, the probability of landing on 2 on any spin is $\frac{1}{3}$.

   She spins the spinner twice.
   Work out the probability that the spinner lands on the side numbered 2 both times.

   .... **(2)**

Guided 2   On any day the probability that Jake is late for school is 0.2.
   On any day the probability that Jake is late home from school is 0.3.
   These events are independent.

   (a) On Monday, what is the probability that Jake is late for school and late home?

   P(Late for school and late home) = 0.2 x .... = ...   **(1)**

   (b) What is the probability that Jake is late for school on Monday and on Tuesday?

   .... **(1)**

   (c) What is the probability that Jake is not late for school and is late home on Monday?

   P(Not late for school) = 1 − .... = ...

   P(Not late for school and late home) = ... x 0.3 = ...   **(2)**

3   The diagram shows two spinners.

   The probability of landing on any section is the same.
   Eric spins each spinner once.
   (a) Work out the probability that both spinners land on the number 1.

   Spinner A   **Spinner B**

   .... **(2)**

   (b) Work out the probability that spinner A lands on the number 1 and spinner B lands on the number 3.

   .... **(2)**

   > The two events, 1 on A and 3 on B, are independent. Take into account that there are two 3s on B.

4   Alex has a biased spinner.
   The spinner can land on the number 1, the number 2, the number 3 or the number 4.
   The table gives the probabilities of the spinner landing on each number.

Number	1	2	3	4
Probability	0.3	0.25	0.2	x

   Alex spins the spinner twice.
   Work out the probability that the spinner lands on the number 4 both times.

   .... **(3)**

Probabilities from tables

1 Mike carried out a survey on whether or not people had used the local gym.

The results are shown in the two-way table.

	Men	Women	Total
Used gym	25	19	44
Not used gym	15	21	36
Total	40	40	80

One person is chosen at random from the survey.

(a) What is the probability that this person is a man? .... (1)

(b) What is the probability that this person has used the gym? .... (1)

(c) What is the probability that this person is a man who has not used the gym? .... (1)

Guided

2 The two-way table gives information about the favourite summer activity of some students.

A student is selected at random.

	Boys	Girls	Total
Swimming	17	13	
Tennis	7	18	
Total			

Find the probability that this student will
(a) be a boy

Total number of boys = ...

Total number of students = ...

> You should always complete the two-way table first.

The probability that the student will be a boy = ... (2)

(b) have tennis as their favourite activity .... (1)

(c) be a boy who has tennis as his favourite activity. .... (1)

3 Here is some information about 50 students at a summer school.
Each student had to choose Art or Dance or Drama.
17 of the students chose Art.
20 of the students were girls. 12 of the girls chose Dance.
5 of the boys chose Drama and 3 of the girls chose Art.

(a) Complete the two-way table to show this information.

	Art	Dance	Drama	Total
Boys				
Girls				
Total				50

(3)

A student is to be selected at random.
(b) Find the probability that this student will be a boy who chose Dance. .... (1)

A student is to be selected at random from the girls.
(c) Find the probability that this girl will have chosen Drama. .... (1)

> The probability will be out of 20 as there are 20 girls.

Experimental probability

1. Araminta has a biased coin. She spins the coin 120 times. She gets heads 54 times.

   Araminta is going to spin the coin one more time. Use her results to estimate the probability of her next spin getting

   > From the experiment, the probability of heads is $\dfrac{54}{120}$

   (a) a head   .... **(1)**

   (b) a tail.   .... **(1)**

2. Alfie, Becky and Colin each roll the same biased dice and count the number of times it lands on the number 1.

   Their results are shown in the table.

Name	Alfie	Becky	Colin
Number of 1s	39	21	42
Number of throws	120	72	108

   (a) Which of the three people has the best data to work out the probability of getting the number 1? Give a reason for your answer.

   .... **(1)**

   Alfie uses his results to estimate the probability of getting the number 1 on one roll.

   (b) Write down Alfie's estimate.   .... **(1)**

   (c) Using the combined data, work out a better estimate for the probability of getting the number 1 on one roll.

       .... **(1)**

3. Zak has a biased coin.

   When the coin is spun once, the probability of getting a head is 0.55.

   > You can estimate the number of times you will get a head by using:
   > probability × number of spins.

   Zak is going to spin the coin 80 times.

   Work out an estimate for the number of heads he will get.   .... **(2)**

> **Guided**

4. Alice has an ordinary dice. She plays a game. Every time she rolls a 3 she gets two sweets. She rolls the dice 60 times. Work out an estimate for the number of sweets she gets.

   > First work out an estimate for the number of 3s she should get.

   Estimate of the number of times she should get a 3 is ...... × .... = ...

   Estimate of the number of sweets she should get is ...... × .... = ... **(2)**

5. This is a fair spinner.

   Every time Rob spins the spinner, he gets the number of sweets that the spinner lands on.

   Rob spins the spinner 60 times.
   Work out an estimate for the number of sweets he should get.   .... **(2)**

Risk

Guided 1 The table shows the number of injuries to ankles, fingers and knees of some players in a cricket season.

	Ankles	Fingers	Knees
Number of injuries	8	15	9
Number of games	40	60	36

Work out the risk of each type of injury.

(a) Ankles $\dfrac{.....}{40}$

(b) Fingers $\dfrac{15}{.....}$

(c) Knees .... **(3)**

2 A company sells driers, washing machines and dishwashers.

The company has information about how many it sold last year and how many of them broke down within the first year of use.
The information is given in the table.

> It can be useful to think of risk as the same as experimental probability.
> There were 4320 driers sold, of which 28 broke down, so the experimental probability of a breakdown can be worked out.

	Driers	Washing machines	Dishwashers
Number of breakdowns	28	45	19
Number sold	4320	6003	3848

(a) Work out the risk of a breakdown in the first year of use for each item.

(i) Drier .... **(1)**

(ii) Washing machine .... **(1)**

(iii) Dishwasher .... **(1)**

This year the company has sold 4580 washing machines.

(b) Work out an estimate for the number of these washing machines that will break down in the first year of use.

.... **(2)**

3 An insurance company insures racing bikes.
Last year the average cost to the company of replacing a racing bike was £3380.

This year the company insures 6580 racing bikes.
The probability that a racing bike will need replacing by the insurance company this year is 0.025.

(a) Work out the total cost of the risk to the insurance company. .... **(2)**

(b) Work out the minimum amount the insurance company should charge to insure each racing bike.

.... **(2)**

Probability trees

Guided 1   There are 6 green beads and 4 blue beads in a bag.

Jim takes a bead from the bag, writes down its colour and replaces it.
Then Hannah takes a bead from the bag.

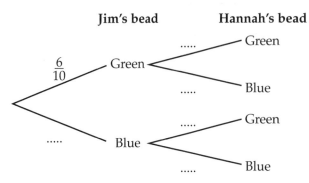

> The sum of the probabilities on each pair of branches must be 1.

(a) Complete the probability tree diagram. **(2)**

(b) Work out the probability that Jim takes a green bead and Hannah also takes a green bead.

$\dfrac{6}{10}$ × .... = ... **(1)**

(c) Work out the probability that they both take beads of the same colour.   .... **(2)**

> You have calculated the probability of *green, green* in part (b) so you only have to find the probability of *blue, blue* and add them together.

2   Every weekday the probability that Nasser wakes up late is 0.1.

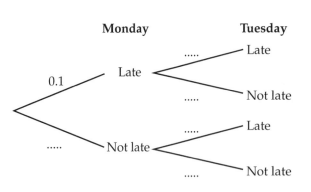

> To work out probabilities in a tree diagram you multiply along the branches and add up the outcomes.

(a) Complete the probability tree diagram for Monday and Tuesday. **(2)**

(b) Work out the probability that Nasser will wake up late on both days.   .... **(1)**

(c) Work out the probability that Nasser will wake up late on just one of the days.

                                                 .... **(1)**

Conditional probability

Aiming higher

Guided

1 Hannah's sock drawer has 6 green socks and 4 blue socks in it.

Hannah randomly takes one sock from the drawer and puts it on. She then takes another sock at random from the drawer.

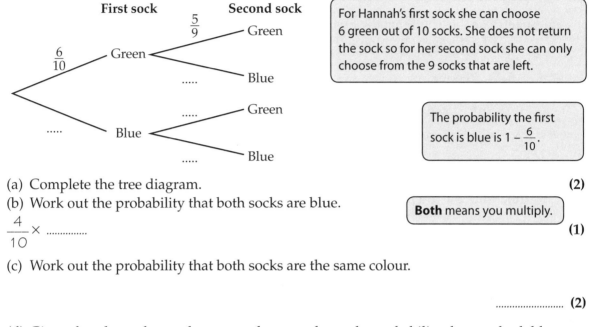

For Hannah's first sock she can choose 6 green out of 10 socks. She does not return the sock so for her second sock she can only choose from the 9 socks that are left.

The probability the first sock is blue is $1 - \frac{6}{10}$.

(a) Complete the tree diagram. (2)

(b) Work out the probability that both socks are blue.

$\frac{4}{10} \times$ .... (1)

Both means you multiply.

(c) Work out the probability that both socks are the same colour.

.... (2)

(d) Given that the socks are the same colour, work out the probability they are both blue.

.... (2)

Aiming higher

2 Here are six cards.

$\boxed{1}$ $\boxed{1}$ $\boxed{1}$ $\boxed{2}$ $\boxed{2}$ $\boxed{3}$

You can sketch a tree diagram to help you think about all the possible outcomes.

Mary takes a card at random and puts it in her pocket.
She then takes a second card at random from the five remaining cards.

(a) Work out the probability that both cards have the number 1 on them.

.... (2)

(b) Work out the probability that both cards have the same number on them.

.... (2)

(c) Work out the probability that the sum of the numbers on the two cards is 4.

.... (2)

Aiming higher

3 There are 8 students in a room. 5 of the students are boys.
3 students are selected at random.

Work out the probability that more girls than boys are selected.

.... (2)

Probability formulae

Aiming higher

1 Beth has a box of scarves.
There are 5 blue scarves and 3 red scarves in the box.

Beth takes a scarf from the box.
She then takes another scarf from the box.

Let X be the event that the first scarf is blue.
Let Y be the event that the second scarf is blue.

First scarf   Second scarf

$\frac{5}{8}$ Blue
     $\frac{4}{7}$ Blue
     $\frac{3}{7}$ Red

$\frac{3}{8}$ Red
     $\frac{5}{7}$ Blue
     $\frac{2}{7}$ Red

(a) Write down the value of $P(Y \mid X)$. .... **(1)**

(b) Work out $P(X \cup Y)$.

.... **(2)**

(c) Work out $P(Y)$.

.... **(2)**

(d) Work out $P(X \cap Y)$.

.... **(2)**

Aiming higher

2 A and B are two events.

Guided

$P(A) = 0.3$, $P(B) = 0.4$, $P(A \cup B) = 0.55$

(a) Work out $P(A \cap B)$.

> Use $P(A \cup B) = P(A) + P(B) - P(A \cap B)$

.... **(2)**

(b) Work out $P(B \mid A)$.

> Use $P(A \cap B) = P(B \mid A) \times P(A)$

.... **(2)**

(c) Are the events A and B independent? Give a reason for your answer.

If A and B are independent, then $P(A \cup B) = P(A) + P(B)$. In this case ....

.... **(2)**

Aiming higher

3 Luka has 12 red cards and 4 blue cards. 6 of the red cards have the letter A on them.
3 of the blue cards have the letter A on them. Luka takes a card at random from the 16 cards.

Let X be the event that he takes a red card.
Let Y be the event that he takes a card with the letter A on it.

Work out

(a) $P(X \mid Y)$ .... **(1)**

(b) $P(X \cap Y)$ .... **(1)**

(c) $P(X \cup Y)$ .... **(1)**

Simulation

Guided

1 The table gives information about how often a train service was late.

Punctuality	On time	Late	Very late
Percentage	78	15	7

> Two-digit random numbers run from 00 to 99 so there are 100 of them.

(a) Show how two-digit random numbers can be used to simulate the punctuality of these trains.

Since the figures are percentages out of 100, on time (78%) can be represented

by the numbers .... , late (15%) by the numbers 78 to ....

and very late by .... **(1)**

(b) The following random numbers have been generated. Use them to perform a simulation of 12 trains.

51   91   94   47   19   56   22   26   79   37   42   15   32   90   47   43   76   30

51 represents 'On time', 91 represents ....

.... **(1)**

(c) In this simulation of 12 trains, in how many cases were consecutive trains late or very late?

.... **(1)**

2 A shop sells ice cream in 3 flavours: vanilla, cherry and chocolate.

Based on previous sales, the shopkeeper knows that out of every 20 customers, 9 choose vanilla, 4 choose cherry and 7 choose strawberry.
Random numbers in this table are going to be used to simulate sales.

> The proportion of the 100 random numbers assigned to cherry must be the same as the proportion of customers who choose cherry.

Flavour	Vanilla	Cherry	Strawberry
Random numbers	00–44	45–64	65–99

(a) Explain why the numbers 45–64 were assigned to cherry.

.... **(1)**

(b) Use the random numbers below to carry out a simulation of 10 sales.

23   09   63   71   90   14   63   43   25   04   60   57   31   64   14   06   76   44

.... **(1)**

(c) Compare the proportion of vanilla sales in the simulation with the proportions previously observed.

.... **(1)**

Probability distributions

Aiming higher

1 The table describes a probability distribution for Y.

Y	1	2	3	4	5
P(Y)	0.3	k	k	k	k

(a) Work out the value of k. .... **(2)**

(b) Find the probability that $Y > 2$. .... **(2)**

Aiming higher

Guided

2 A box has eight counters numbered 1 to 8.

A counter is selected at random from the box. The counter is then replaced.
Let X be the number on the selected counter.

(a) Describe the probability distribution of X.

The distribution is uniform with all probabilities .... **(1)**

(b) Find the probability that $X < 4$.

Probability of taking each counter $= \dfrac{1}{8}$ | There are three values less than 4.

Probability that $X < 4 =$ .... **(1)**

The selection process above is repeated 20 times.

(c) Work out an estimate for how many times the number 8 is selected. .... **(1)**

| Use 20 multiplied by the probability that $X = 8$.

Aiming higher

3 Four people select one of the digits from 0 to 9 at random.
X is the digit written down by the first person.

(a) Describe the probability distribution of X. .... **(1)**

(b) Find the probability that $X < 7$. .... **(2)**

(c) Find the probability that all four people write down a digit less than 7. .... **(2)**

Aiming higher

4 Harry has a biased spinner. He spins the spinner once.
The table describes the probability distribution of X, the number the spinner lands on.

X	1	2	3	4	5
P(X)	k	0.3	0.3	0.3	k

(a) Find the value of k. .... **(2)**

(b) Find the probability that $X > 2$. .... **(2)**

The same spinner is spun twice.

(c) Work out the probability that it lands on a number greater than 2 exactly once.

.... **(2)**

The binomial distribution 1

1 Here is a spinner.

Each time the spinner is spun, the probability of the spinner landing on red is twice the probability of the spinner landing on blue.

Jean spins the spinner twice.

Spinner

Blue

Red

(a) Calculate the probability that she lands on red both times.

The probability of red on any spin is $\dfrac{\cdots}{3}$

So the probability of two reds is $\left(\dfrac{\cdots}{\cdots}\right)^2 =$ .... **(2)**

(b) Calculate the probability that the spinner lands on red once and blue once, in either order.

> You have to consider red then blue as well as blue then red.

.... **(2)**

2 The probability that Gareth's train arrives late on any day is 0.1.

(a) Work out the probability that his train will be late both on Monday and on Tuesday.

.... **(2)**

(b) Work out the probability that his train will be on time both on Thursday and on Friday.

.... **(2)**

3 Erica has a biased coin.
When she spins the coin twice she knows that the probability she gets heads both times is $\dfrac{1}{16}$.

> If p is the probability of getting a head on a single throw then p^2 is the probability of getting heads on two throws.

Erica spins the coin twice.

(a) Calculate the probability that she gets exactly 1 head.

.... **(3)**

(b) Write down one assumption that you made to find the answer to part (a).

.... **(1)**

Had a look ☐ Nearly there ☐ Nailed it! ☐

The binomial distribution 2

Aiming higher

1 A spinner has five equal segments: 3 coloured blue and 2 coloured red.
The spinner is spun three times.

Let Y be the number of times the spinner lands on blue.
(a) Describe the probability distribution that is a suitable model for Y.

.... **(1)**

(b) Work out $P(Y = 1)$.

.... **(2)**

(c) Work out the probability that the spinner lands on blue twice.
You may use $(p + q)^3 = p^3 + 3p^2q + 3pq^2 + q^3$

> You will always be told the relevant expansion in the exam.

.... **(1)**

Aiming higher

Guided

2 Here are six cards.

Jenny takes a card at random, notes its number and replaces it.

| 1 | 1 | 1 | 2 | 2 | 3 |

She does this four times.
Let X be the number of times Jenny takes the number 3.

(a) Describe the probability distribution that is a suitable model for X.

.... **(1)**

(b) Find $P(X = 2)$.
You may use $(p + q)^4 = p^4 + 4p^3q + 6p^2q^2 + 4pq^3 + q^4$

$6p^2q^2$ gives .... **(2)**

(c) Find $P(X > 2)$.

The terms $p^4 + 4p^3q$ ....

> You need the terms that correspond to values $X = 3$ and $X = 4$.

(2)

Aiming higher

3 Here are five tiles.

The tiles are placed in a bag and one is taken out at random.
The tile is replaced.
This process is done five times.

| + | + | + | O | O |

Let Y be the number of tiles taken with O on.

(a) Describe the probability distribution that is a suitable model for Y.

.... **(1)**

(b) Work out the probability of getting exactly three tiles with O on.

.... **(1)**

(c) Work out the probability of getting more tiles with O on than + on.

.... **(1)**

You may use $(p + q)^5 = p^5 + 5p^4q + 10p^3q^2 + 10p^2q^3 + 5pq^4 + q^5$

The normal distribution

1 These normal distribution graphs give information about the marks students got in two tests.

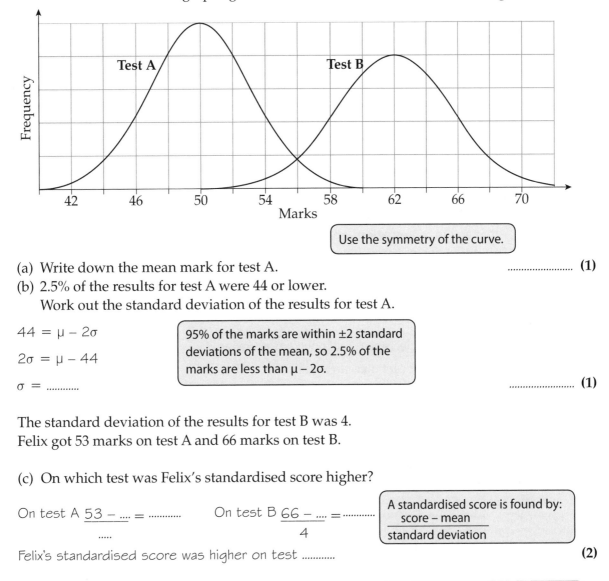

Use the symmetry of the curve.

(a) Write down the mean mark for test A.   .... **(1)**

(b) 2.5% of the results for test A were 44 or lower.
    Work out the standard deviation of the results for test A.

$44 = \mu - 2\sigma$

$2\sigma = \mu - 44$

| 95% of the marks are within ±2 standard deviations of the mean, so 2.5% of the marks are less than $\mu - 2\sigma$. |

$\sigma = ...$ .... **(1)**

The standard deviation of the results for test B was 4.
Felix got 53 marks on test A and 66 marks on test B.

(c) On which test was Felix's standardised score higher?

On test A $\dfrac{53 - ....}{.....} = ...$ On test B $\dfrac{66 - ....}{4} = ...$

A standardised score is found by:
$\dfrac{\text{score} - \text{mean}}{\text{standard deviation}}$

Felix's standardised score was higher on test ....   **(2)**

2 The heights of 1000 adult males have a normal
distribution with almost all of the heights
symmetrically placed between 162 cm and 190 cm.

A normal distribution is always
symmetrical about the mean.
Almost all the heights are within
3 standard deviations of the mean.

(a) Estimate the mean and the standard deviation.

.... and .... **(2)**

(b) The height exceeded by 975 of the males is x cm. Find the value of x.   .... **(1)**

3 The lengths of wheat plants are normally distributed.
In a large sample of wheat plants 50% of the plants had a height less than 75 cm
and 97.5% had a height of at least 72 cm.

(a) Estimate the mean and the standard deviation.   .... and .... **(2)**

There were 800 wheat plants in a sample.
(b) Estimate the number of these plants with heights between 75 cm and 78 cm.

.... **(2)**

Quality control 1

Aiming higher

Guided

1 A production line makes precision bearings with a target diameter of 2.0 cm.

Tests take place every 10 minutes, when eight bearings are selected and measured. From previous experience, the standard deviation of the sample means is 0.04 cm.

> $\mu \pm 2\sigma$ for the warning limits.
> $\mu \pm 3\sigma$ for the action limits.

(a) Work out suitable warning and action limits for the sample mean.

....

.... **(2)**

The table shows the diameters in one sample.

Diameter (cm)	1.96	2.05	1.99	2.0	2.04	2.13	2.20	2.01

(b) What action, if any, should be taken in the light of this sample? You must give a reason for your answer.

The mean of this sample is .... This value is .... , so the action that ....

.... **(2)**

Aiming higher

2 A company makes needles for science laboratories. The stated length of each needle is 10.5 mm.

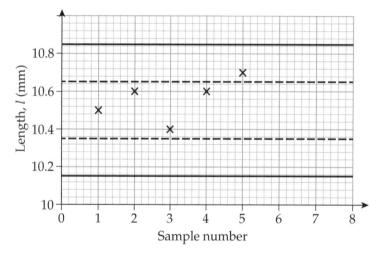

Sample number

The grid shows the quality control chart for the process.

(a) Write down the target length. .... **(1)**

A sample of 10 needles is taken every 10 minutes and the mean is plotted on the chart.

(b) Describe what action, if any, should have been taken for the first five samples.

.... **(1)**

Sample 6 has a mean of 10.9 mm.

(c) Describe what action, if any, should be taken.

.... **(1)**

Quality control 2

Aiming higher

Guided

1 A factory uses the control chart shown to check on the variability in thickness of the glass it produces.

In this case there are no lower limits for the range.

(a) Explain why the range has been chosen, rather than the mean.

The mean is a measure of ....

The range is a measure of ....

.... **(1)**

(b) What should have happened as a result of sample 3?

The range for sample 3 falls

in between the warning and the action limits so

.... **(1)**

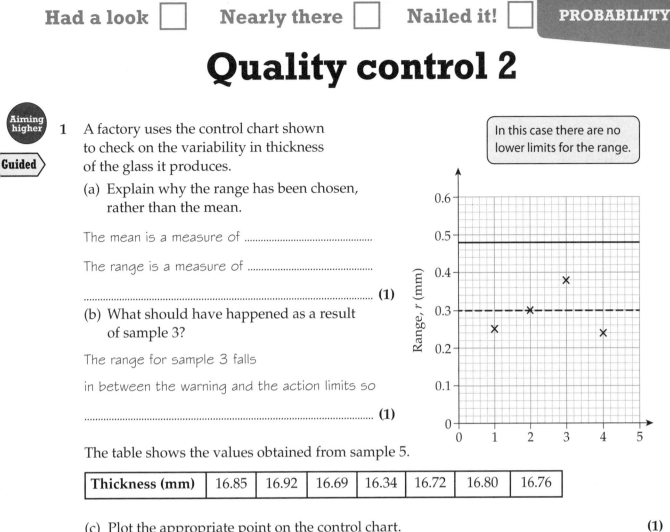

The table shows the values obtained from sample 5.

Thickness (mm)	16.85	16.92	16.69	16.34	16.72	16.80	16.76

(c) Plot the appropriate point on the control chart. **(1)**

(d) What action, if any, should be taken in the light of this sample? You must give a reason for your answer.

.... **(2)**

Aiming higher

2 A machine is used to fill perfume bottles.

The production process is checked by taking samples of 10 bottles and plotting the sample median.

You can solve problems involving the sample median in the same way as problems involving the sample mean.

The control chart shows the warning and action limits.

(a) What volume should be shown on the bottle as the minimum contents? You must give a reason for your answer.

.... **(1)**

(b) What action, if any, should have been taken for each of the five samples shown?

.... **(1)**

Answers

The number given to each topic refers to its page number.

1. Types of data
1. (b) and (c)
2. (b) and (d)
3. (a) Qualitative   (b) Quantitative   (c) Qualitative
   (d) Quantitative
4. (a) Discrete (b) Discrete (c) Continuous
5. (a) Qualitative   (b) Quantitative, discrete   (c) Rank
6. For some people it may be difficult to decide on what their eye colour is.

2. Measurements and variables
1. (a) Explanatory – rainfall, response – number of flowers
   (b) Explanatory – age, response – length of arm
   (c) Explanatory – number of rooms, response – price
   (d) Explanatory – hours or revision, response – exam grade
2. (a) (i) 23.5 cm   (ii) 24.5 cm
   (b) (i) 241.5 kg   (ii) 242.5 kg
   (c) (i) 1 h 59 m 30 s   (ii) 2 h 0 m 30 s
3. (a) Continuous   (b) Discrete
   (c) Number of characters   (d) Time it took to read
   (e) 18.5 s

3 Sampling frames, pre-tests and pilots
1. (a) Biased in favour of people who do not have a full-time job.
   (b) He will not get any families who have no children.
   (c) Biased against people who do not have a home phone or who are ex-directory.
   (d) Biased in favour of those who already use the library.
   (e) Biased in favour of those who have access to emails.
2. (a) A numbered list of all year 11 students
   (b) A numbered list of all customers from the previous month
   (c) A list of all the householders in the estate
   (d) A list of all dwellings from which waste is collected
3. Biased towards people who are already likely to be in favour of organic foods.
4. (a) Advantage – complete information about the population
   Disadvantage – can take a long time, can be very expensive
   (b) Advantage – can be done quicker than a census
   Disadvantage – some groups may be overlooked, estimates of the population will not necessarily be accurate
5. 38 41 03 48 34 07 29 13 31 10

4. Experiments and hypotheses
1. Bigger gardens have a greater number of different vegetables in them.
2. Salty chips taste better than non-salty chips.
3. (a) Two groups of seeds – one with no watering and one with watering
   (b) Water with no copper sulphate added
   (c) A group of people drinking water before bedtime
4. (a) Using slug pellets will get rid of slugs.
   (b) Have two patches of garden. Treat one with slug pellets and leave the other alone. Count the number of slugs in each patch.
5. 150
6. 576

5. Stratified sampling
1. 12
2. (a) 250   (b) 6
3. (a) There are about twice as many men in full-time work in the population as there are women in full-time work in the population.
   (b) 10

6. Further stratified sampling
1. (a) Older people are more likely to object to noise. Younger people are more likely to want improved transport links.
   (b) 28
2. 56

3. 12.7, 16.7, 20.7, rounded to 12, 17 and 21 for a sample of size 50

7. Further sampling methods
1. (a) Systematic
   (b) Advantage – straightforward to do, good coverage of the street
   Disadvantage – results may be affected by patterns in the types of houses in the street (e.g. flats vs detached houses with children)
2. (a) Quota
   (b) Advantage – easy to get the required proportions of people
   Disadvantage – may not be as representative of the population as random sampling would be
3. In cluster sampling the cluster is treated as the unit and all members of the cluster are sampled. In stratified sampling a random sample of a suitable size is taken from each stratum of the population.
4. (a) If the cars are passing quickly he may not be able to record successive cars (which could feature in a random sample). Systematic sampling would allow Alex the time to record his observations.
   (b) If Alex allows 10 seconds per car recorded then a total of 500 seconds is needed.
   (c) It is likely that his sample will be biased towards people going to and coming from work.
5. (a) Because the staff are widespread it would take a lot of travelling time to interview them all.
   (b) Use cluster sampling. Select 5 or 6 police headquarters at random and interview all the people at those selected headquarters.

8. Sampling overview
1. (a) Census   (b) Cluster   (c) Quota
   (d) Systematic   (e) Quota
2. (a) It takes a long time to analyse the data; it is very expensive; it is very time consuming to prepare.
   (b) So that reliable information can be collected from the whole population.
   (c) Already in electronic form so easier to analyse; easier to check who has completed the census form.

9. Data capture sheets
1.

Gender	Tally	Frequency
Male		
Female		

2.

Type of plant	Tally	Frequency
Crocus	| | |	3
Hyacinth	++++ | | |	8
Daffodil	++++ | |	7
Tulip	++++ | |	7

3. (a)

Type	Tally	Frequency
Adult swimmer	++++	5
Adult non-swimmer	++++	5
Child swimmer	++++ | |	7
Child non-swimmer	| | |	3

(b) Child swimmer

10. Interviews and questionnaires
1. (a) It is too open and would allow too wide a variety of responses.
   (b) It is a leading question.
2. (a) Questions can be explained or responses can be followed up immediately to clarify.
   (b) Interviews take a long time to do.
   There is less likelihood of young people giving completely honest responses.

(c) It is too open and would allow too wide a variety of responses.

(d) To check questions can be understood.
To estimate the time taken.

3. (a) Many illnesses and treatments can be very complex so it may need a doctor to explain things.

(b) People might not remember or may only remember the first time the illness was definitely diagnosed.

11. Questionnaires

1. How many times do you shop at this supermarket each week?

Never   Once   Twice   More than twice

2. There is no time frame – is the use per day or per week?
The response boxes overlap, e.g. at 3 hours.

3. (a) There is no time frame – is the use per lesson or per day or per week?
It is not exhaustive – there is no option for not sending any text messages.

(b) 'How many text messages did you send yesterday?'

☐   ☐   ☐   ☐

0 to 10   11 to 2   21 to 30   More than 30

4. How far do you usually travel to away matches?

☐   ☐   ☐   ☐   ☐

0–50   51–100   101–150   151–200   More than 200
miles   miles   miles   miles   miles

12. Capture/recapture

1. (a) 0.24   (b) 375

(c) That marking a fish does not change the chance it will be caught.

2. 555

3. (a) 1792   (b) 2400

(c) It could be due to sampling error or to migration of extra barnacle geese into the population.

4. (a) The number caught that are marked is too small.
A change of 1 would change the estimate by a huge amount.

(b) 1000

13. Frequency tables

1 (a) 16   (b) 41   (c) Aged 61 or over

2. (a)

Age, T seconds	Tally	Frequency
$0 < T \leq 10$	⊦⊦⊦⊦	5
$10 < T \leq 20$	⊦⊦⊦⊦ ‖	7
$20 < T \leq 30$	⊦⊦⊦⊦ ‖‖	9
$30 < T \leq 40$	‖‖	4

(b) 13

3 (a) 4   (b) 43   (c) 12 + 26 + 21 + 12 = 71

14. Two-way tables

1. (a)

	Had a cold	Not had a cold	Total
Men	38	43	(81)
Women	20	(36)	56
Total	(58)	(79)	(137)

(b) 81   (c) 21

2. (a)

	Drums	Guitar	Vocals	Total
Girls	0	6	15	21
Boys	12	12	7	31
Total	12	18	22	52

(b) 12

3. 16

	Walk	Bus	Other	Total
Girls	14	8	4	26
Boys	20	8	6	34
Total	34	16	10	60

15. Pictograms

1. (a) 40   (b) 50   (c)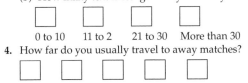

2.

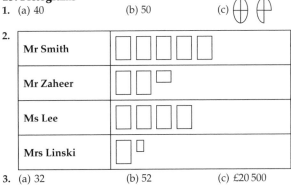

Mr Smith	
Mr Zaheer	
Ms Lee	
Mrs Linski	

3. (a) 32   (b) 52   (c) £20 500

16. Bar charts and vertical line graphs

1. The vertical scale does not have equal spacing.
One label is missing from the horizontal axis.
The vertical axis does not have a label.
The bars do not have equal widths.

2. (a)

Fruit	Tally	Frequency
Apple	‖‖	4
Banana	⊦⊦⊦⊦ ‖	7
Orange	‖‖	4
Peach	‖‖	3
Pineapple	‖	2

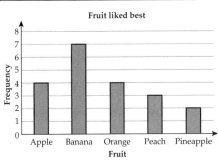

3. The profit axis does not start from zero, so it makes it look as though company B made more than twice as much profit as company C.

17. Stem and leaf diagrams

1. (a)

0	0 0 2 4 4 7 8
1	0 2 4 4 5 7 7
2	1 2 3 7 8
3	0 1

Key: 3 | 1 means 31 orchids

(b) $\frac{7}{21} = \frac{1}{3}$

2. (a)

0	6 7 8
1	2 2 3 3
2	0 6 7 8
3	3 4 6

Key: 3 | 4 means 34 moths

(b) $\frac{9}{14}$

3. (a)

12	6 8 9
13	2 7 8 8
14	5 9 9
15	3 5 6 6 7 7
16	5
17	0 1 5

Key: 17 | 5 means 175 words

(b) 20%

18. Pie charts

1. (a) $\frac{15}{360}$ (b) £8444

2. 16

19. Drawing pie charts

1. Pie chart with the following angles:

Regular walking	Cycling	Visiting the gym	Swimming	Other
120°	90°	60°	50°	40°

2. Pie chart with the following angles:

Europe	America	Africa	Asia
117°	150°	45°	48°

3. Pie chart with the following angles (to the nearest degree):

Cereals	Green crops	Grass	Set aside
119°	37°	151°	54°

20. Bar charts

1. (a) Adult attendance fell each week.
   (b) 3rd week   (c) 31

2.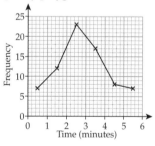

3. (a) 30%   (b) 40%   (c) 90

21. Pie charts with percentages

1. (a) Pie charts with the following angles (to the nearest degree):

	Service	Retail	Other
1980	83°	133°	144°
2010	140°	115°	104°

   (b) You cannot tell because the total numbers working in 2010 and 1980 are not known.

2. (a) Law 80, Science 75, Maths 60, Other 35
   (b) No, because the total number does not change nor the number who chose Maths.

22. Using comparative pie charts

1. (a) 784   (b) 436   (c) School B
2. 7.5 cm

23. Frequency polygons

1. (a)

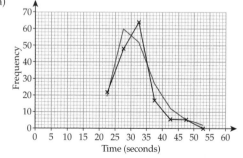

   (b) The peak of waiting times occurs between 2 and 3 minutes
2. (a)

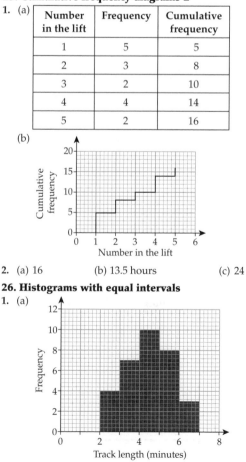

   (b) Both distributions have a single peak.
   The peak time for women was between 25 and 30 seconds.
   The peak time for men was between 30 and 35 seconds.
   There were more women who could hold their breath for more than 40 seconds than men who could hold their breath for more than 40 seconds.

24. Cumulative frequency diagrams 1

1. (a)

Wage, W (£)	0 < W ≤ 200	0 < W ≤ 400	0 < W ≤ 600	0 < W ≤ 800	0 < W ≤ 1000
Frequency	10	25	43	58	67

   (b)

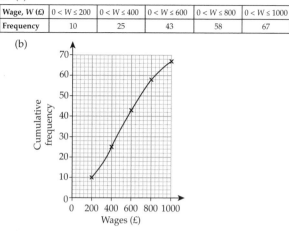

2. (a)

Number N per m²	0 < N ≤ 10	0 < N ≤ 20	0 < N ≤ 30	0 < N ≤ 40	0 < N ≤ 50	0 < N ≤ 60
Cumulative frequency	13	20	35	50	62	70

   (b)

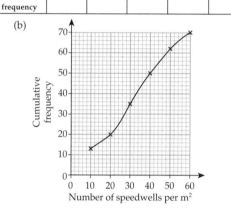

25. Cumulative frequency diagrams 2

1. (a)

Number in the lift	Frequency	Cumulative frequency
1	5	5
2	3	8
3	2	10
4	4	14
5	2	16

   (b)

2. (a) 16   (b) 13.5 hours   (c) 24

26. Histograms with equal intervals

1. (a)

   (b) $\frac{11}{32}$

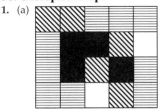

2. (a) 12
   (b) 25
   (c) 6

27. Histograms with equal intervals

1. (a)

Area, A (m²)	Frequency	Frequency density
$0 < A \leqslant 10$	8	0.8
$10 < A \leqslant 15$	13	2.4
$15 < A \leqslant 20$	22	4.4
$20 < A \leqslant 40$	34	1.7
$40 < A \leqslant 100$	18	0.3

(b)

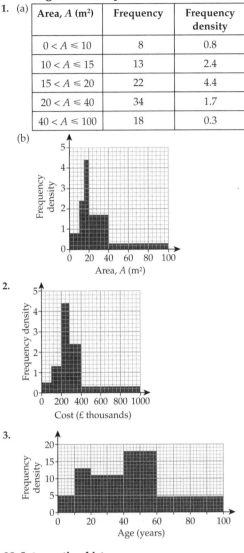

2.

3.

28. Interpreting histograms

1. $\frac{272}{564}$ or 48.2%

2. (a) 36   (b) 31

29. Population pyramids

1. (a) Looking at the overall shape, the bars in the UK look about the same for men and women, which means that there are the same numbers of men and women in each age group.
   In Iran the bar for over 60 for women is longer than the bar for over 60 for men, meaning that a greater percentage of the women live longer compared with the men.
   In the UK there is a greater percentage of both men and women who are older.

2. (a)

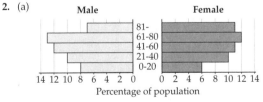

(b) The percentages of men and women are roughly the same except in the oldest age group.

30. Choropleth maps

1. (a)

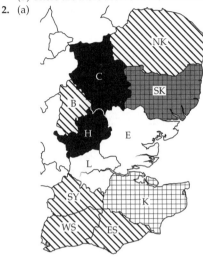

(b) There are more acorns in the middle of the woodland.

2. (a)

(b) The rainfall gets lower towards the South East.

31. Mode, median and mean

1. (a) 5   (b) 6   (c) 5.8

2. (a) 29.8   (b) 32
   (c) (i) Mean will increase.
       (ii) Median will stay the same as the order of the marks will not change.

3. 1.6

4. 9

32. Mean from a frequency table

1. 4

2. 3.65

3. 1.9 (1 d.p.)

33. Mean from a grouped frequency table

1. (a) 8.2 km (1 d.p.)
   (b) Because midpoints of intervals have been used, not the actual values themselves.

2. £7.39

3. 4.3 hours (1 d.p.)

34. Mode and median from a frequency table

1. (a) 3   (b) 4

2. (a) 8   (b) 7
   (c) (i) Both 7 and 8 are modes.
       (ii) Median is unchanged.

35. Averages from grouped frequency tables

1. (a) $6 < L \leqslant 8$ (b) $4 < L \leqslant 6$ (c) 5.70

2. (a) $2 < T \leqslant 3$ (b) $3 < T \leqslant 4$
   (c) $3 < T \leqslant 4$
   (d) The 49th data value is now in that interval.

36. Which average?

1. (a) £215   (b) £237
   (c) Not the mode as there are two values which each occur twice. Not the mean because it is affected by the large value of £420. So the median.

2. (a) 7   (b) 7.35
   (c) (i) The median stays the same.
       (ii) The mean will be reduced.

3. (a) 205 g   (b) Median will be reduced to 202.5 g.

37. Estimating the median

1. (a) 34, 43, 47
   (b)

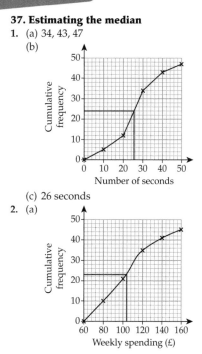

   (c) 26 seconds

2. (a)

   (b) £104
   (c) (i) Median rises to £106.
       (ii) As can be seen by reading off the diagram at 25.5.

38. The mean of combined samples

1. 169.2 cm
2. 6.1 (1 d.p.)
3. 2.25
4. 72.15 mm
5. 59%

39. Weighted means

1. (a) No, because her weighted mean mark is 63.
   (b) 63
2. 32.8
3. (a) 7.875   (b) 7

40. Measures of spread

1. 8
2. (a) 9°C   (b) 5°C or −8°C
3. Boys' range = 10, girls' range = 10. James is not correct.
4. (a) 1   (b) 7   (c) 6
5. 3

41. Box plots

1. (a) Median = 20, Q_1 = 13, Q_3 = 31
   (b)

2. (a) Median = 18, Q_1 = 13, Q_3 = 24
   (b)

   (c)

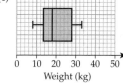

42. Interquartile range and continuous data

1. (a) 6, 31, 58, 77, 87, 94, 97, 98, 100
   (b)   (c) £315

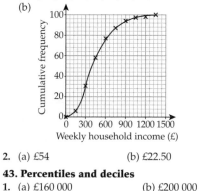

2. (a) £54   (b) £22.50

43. Percentiles and deciles

1. (a) £160 000   (b) £200 000
2. (a) 3.35 hours (3 h 21 m)   (b) 4 hours
   (c) 4.4 hours (4 h 24 m)
   (d) approx 1h 30m

44. Comparing discrete distributions

1. (a) 6.5   (b) 10
   (c) The median for the women is 5 which is less than that of the men so on average the women had used their cars less. The range for the women is 12 so there is greater variation in use by the women than the men, for whom the range is 10.

2. The median for the men (17) is less than the median of the women so on average the men watched fewer television programmes than the women.
   The interquartile range (IQR) for the men was 15 which is greater than the IQR for the women so the central 50% of the women are more clustered about the median

3. (a)

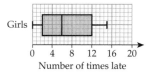

   (b) The median for the boys (7) is greater than that for the girls (6) so on average the boys were late more often.
   The interquartile range (IQR) of the boys (6) is less than the IQR of the girls (10) so the central 50% of the boys are more clustered about the median.

45. Cumulative frequency diagrams and box plots

1.

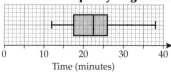

2. (a)

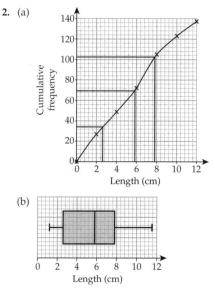

   (b)

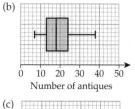

46. Using cumulative frequency diagrams and box plots

1. The median time for the men was 21 minutes, so on average their times were less than those of the women whose median was 26 minutes
   The range was 29 minutes for the men and 37 minutes for the women, so the spread of times was greater for the women. The interquartile range was 9 minutes for the men and 18 minutes for the women, so the central 50% of times for the men were more clustered about the median.

2. (a)

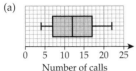

Number of calls

   (b) The median for the boys (9) is lower than the median for the girls (12), so the girls made more calls on average. The interquartile range (IQR) for the boys (6) was less than the IQR for the girls (10) so the central 50% of the boys were more clustered about the median. The range for the boys (17) was less than the range for the girls (18) so the dispersion was less for the boys.

47. Box plots and outliers

1. (a) IQR = 11 − 5 = 6, Q_3 + 1.5 × 6 = 20, so 21 is an outlier.
   (b)

Number of tracks

2. (a) 9   (b) 4
   (c) Girls: IQR = 7, 1.5 × 7 = 10.5, so any values from −5.5 to 22.5 are not outliers.
   Boys: IQR = 4, 1.5 × 4 = 6, 8 + 6 = 14, so 15 is an outlier.

3. (a) Neither group A nor group B have outliers. For A, outliers would be below 2.5 or above 30.5. For B, outliers would be below 3.75 or above 25.75.
   (b) The median of group B (15) is slightly larger than the median (14) of group A so there is no evidence that on average the fertiliser improved growth.
   The range and the IQR are bigger for group A, showing that there was a bigger dispersion in group A, so some plants did very well compared to group B, but others did poorly.

48. Box plots and skewness

1. (a)

   Number of emails

   (b) $Q_3 − Q_2$ = 4 , $Q_2 − Q_1$ = 2. There is some evidence for a (small) positive skew.

2. (a) 8   (b) 4
   (c) The girls' distribution has slight positive skew. The boys' distribution has negative scew.

3. (a)

   Market

   Mass, m (grams)

   (b) 56 g
   (c) The median mass from the market (150 g) was greater than the median mass from the shop (125 g) so on average the onions from the market were heavier.
   The IQR from the shop (56 g) was greater than the IQR from the market (45 g) showing that the central 50% were more clustered in the market.
   The distributions from both the shop and the market have a negative skew, the shop more so.

49. Variance and standard deviation

1. (a) 13.25   (b) 3.64 (3 s.f.)
2. (a) 28.4   (b) 10.5
3. (a) 63   (b) 505   (c) 3.81 (3 s.f.)
4. (a) 5.29   (b) 204.32
5. (a) 50 cm   (b) 22.4 cm (3 s.f.)
   (c) 46.7 cm (3 s.f.)   (d) 21.7 cm (3 s.f.)

50. Standard deviation from frequency tables

1. (a) 2040, 52 900   (b) £19 057   (c) £5904
2. (a) 27, 40.5, 9   (b) 2 kg   (c) 1.08 kg
3. 9.4

51. Simple index numbers

1. (a) 122, 121
   (b) The index number of the adults is slightly greater than the index number for the 16–17 year olds, showing that the adults had a proportionally greater increase in their rate.
2. (a) £1.638 million   (b) 65
3. (a) £509.30
   (b) Yes, his pay should have been £759.32.

52. Chain base index numbers

1. (a) 102.9, 104.6, 103.5
   (b) 2010 as the chain base index number is the largest.
2. (a) 108.3, 107.7
   (b) As the chain base compares year on year increases it can give a misleading impression.

53. Weighted index numbers

1. (a) 122.7
   (b) Change in weighted index = 22.7 which is greater than the change in the CPI.
2. 91.1

54. Standardised scores

1. (a) (i) 1   (ii) −0.25
   (b) (i) 55   (ii) 46
   (c) Vanessa did better than Flora on the numerical test but worse than Flora on the verbal test.
2. Abstract standardised score is 0.53, symbol sorting standardised score is 0.3, so Lee did better on the abstract test.

55. Scatter diagrams and correlation

1. (a) Strong positive   (b) Weak negative   (c) Strong negative
2. (a) Engine size
   (b)

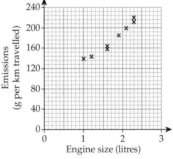

   (c) Strong positive
   (d) As the engine size increases so do the emissions per km travelled.

56. Lines of best fit

1. (a), (c)

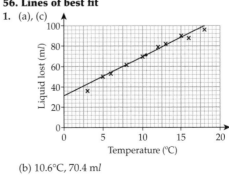

   (b) 10.6°C, 70.4 ml
   (d) 65 ml

2. (a) 163.5, 61.8

(b)

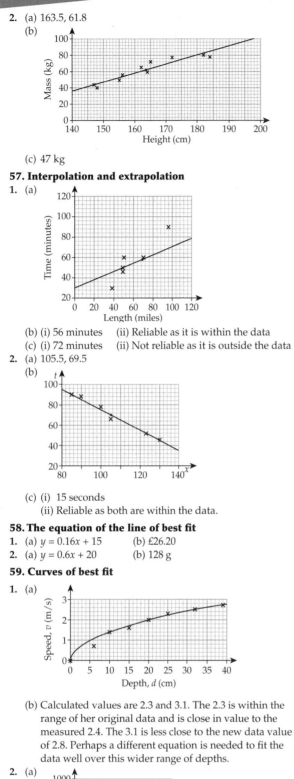

(c) 47 kg

57. Interpolation and extrapolation

1. (a)

(b) (i) 56 minutes   (ii) Reliable as it is within the data

(c) (i) 72 minutes   (ii) Not reliable as it is outside the data

2. (a) 105.5, 69.5

(b)

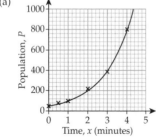

(c) (i) 15 seconds

(ii) Reliable as both are within the data.

58. The equation of the line of best fit

1. (a) $y = 0.16x + 15$ (b) £26.20

2. (a) $y = 0.6x + 20$ (b) 128 g

59. Curves of best fit

1. (a)

(b) Calculated values are 2.3 and 3.1. The 2.3 is within the range of her original data and is close in value to the measured 2.4. The 3.1 is less close to the new data value of 2.8. Perhaps a different equation is needed to fit the data well over this wider range of depths.

2. (a)

(b) (i) 4.3 minutes

(ii) It is outside the range of the original data so is not a reliable answer.

60. Spearman's rank correlation coefficient 1

1. (a) 0.43

(b) There is a positive correlation between the judges; there is a degree of agreement between them.

2. (a) 0.57

(b) There is a positive correlation between the performances on the two tests so the designer's hypothesis is supported.

3. (a) 0.48   (b) −0.48

61. Spearman's rank correlation coefficient 2

1. (a) 0.89

(b) Yes, as the correlation coefficient is close to 1 and the points on the graph are nearly in a straight line.

2. (a) 0.93

(b) Because the correlation coefficient is so high there is support for the hypothesis.

(c) No difference

62. Time series

1. (a) 1500   (b) 38.9°C   (c) 0700 to 0900

2. (a)

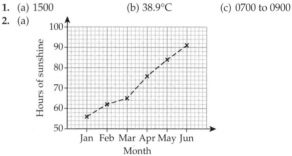

(b) The hours of sunshine start at 56 in January and then they increase.

63. Moving averages

1. (a) 31, 30, 29, 29, 29, 29, 29 (b) The sales stay about the same.

2. (a) 138, 140.   (b) The trend is increasing.

64. Moving averages and trend lines

1. (a)

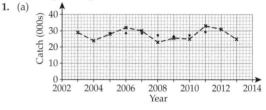

(b) The moving averages remain fairly similar over the period observed. Between 2009 and 2011 they increased.

2. (a)–(c)

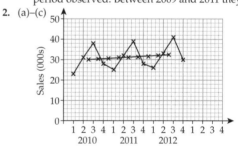

(d) The trend in sales is increasing.

65. The equation of the trend line

1. $y = 0.75x − 1464$

2. $y = −0.27x + 12$

66. Prediction

1. (a) 6.5°C   (b) 16.5°C

(c) Prediction does depend on temperature trend being the same and seasonal variation following the same cycle.

2. (a) 9 000   (b) 61 000   (c) 40 000

(d) The prediction for summer 2007 because it is closer in time to the the data.

67. Probability

1. Head, tail

2. 1, 2, 3, 4, 5, 6

3. (a) 1, 2, 3, 4, 5, 6, 7, 8   (b) 2, 4, 6, 8   (c) 4, 5, 6, 7, 8

4. (a) Likely   (b) Unlikely   (c) Unlikely

5.

68. Sample spaces
1. Head, Tail
2. Head Head, Head Tail, Tail Head, Tail, Tail
3. H1, H2, H3, H4, H5, H6, T1, T2, T3, T4, T5, T6, where H = Head and T = Tail
4. (a) HHT, HTH, THH, HTT, THT, TTH, TTT
   (b) No. There is only one outcome for 3 heads, but three for two heads and a tail.
5. (a) (1, 1) (1, 2) (1, 3) (1, 4), (2, 1) (2, 2) (2, 3) (2, 4), (3 ,1) (3, 2) (3, 3) (3, 4), (4, 1) (4, 2) (4, 3) (4, 4)
   (b) 2   (c) 5
6. (a) (1, 1) (1, 2) (1, 3) (1, 3), (2, 1) (2, 2) (2, 3) (2, 3), (3, 1) (3, 2) (3, 3) (3, 3), (3, 1) (3, 2) (3, 3) (3, 3)
   (b) 2   (c) 2

69. Probability and sample spaces
1. (a) G, G, G, W, W, R, R, R, R
      where G = green, W = white, R = red.
   (b) (i) $\frac{3}{9}$ (ii) $\frac{4}{9}$ (iii) $\frac{6}{9}$
2. (a) (1, 1) (1, 2) (1, 3) (1, 4), (2, 1) (2, 2) (2, 3) (2, 4), (3, 1) (3, 2) (3, 3) (3, 4), (4, 1) (4, 2) (4, 3) (4, 4)
   (b) $\frac{4}{16}$ (c) $\frac{3}{16}$
3. (a) (1, 1), (1, 2) (1, 3) (1, 4) (1, 5) (1, 6) (2, 1), (2, 2) (2, 3) (2, 4) (2, 5) (2, 6) (3, 1), (3, 2) (3, 3) (3, 4) (3, 5) (3, 6) (4, 1), (4, 2) (4, 3) (4, 4) (4, 5) (4, 6) (5, 1), (5, 2) (5, 3) (5, 4) (5, 5) (5, 6) (6, 1), (6, 2) (6, 3) (6, 4) (6, 5) (6, 6)
   (b) $\frac{1}{36}$ (c) $\frac{10}{36}$
4. (a) (2, 2) (2, 3) (2, 4) (3, 3) (3, 2) (3, 4) (4, 2) (4, 3) (4, 4)
   (b) $\frac{3}{9}$ (c) $\frac{3}{9}$

70. Venn diagrams and probability
1. (a) 35   (b) 5   (c) 10
2. (a)

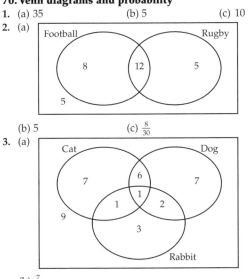

   (b) 5   (c) $\frac{8}{30}$
3. (a)

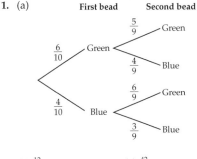

   (b) $\frac{7}{36}$

71. Mutually exclusive events
1. 0.8
2. 0.6
3. (a) 0.25   (b) 0.5
4. (a) (i) 0.55 (ii) 0.75 (iii) 0.7
   (b) (i) 0.64 or $\frac{16}{25}$ (ii) 0.6 or $\frac{15}{25}$ (iii) 0.76 or $\frac{19}{25}$

72. Independent events
1. $\frac{1}{9}$
2. (a) 0.06   (b) 0.04   (c) 0.24
3. (a) $\frac{1}{16}$ (b) $\frac{2}{16}$
4. 0.0625

73. Probabilities from tables
1. (a) $\frac{40}{80}$ (b) $\frac{44}{80}$ (c) $\frac{15}{80}$
2.

	Boys	Girls	Total
Swimming	17	13	30
Tennis	7	18	25
Total	24	31	55

   (a) $\frac{24}{55}$ (b) $\frac{25}{55}$ (c) $\frac{7}{55}$

3. (a)

	Art	Dance	Drama	Total
Boys	14	11	5	30
Girls	3	12	5	20
Total	17	23	10	50

   (b) $\frac{11}{50}$ (c) $\frac{5}{20}$

74. Experimental probability
1. (a) $\frac{54}{120}$ (b) $\frac{66}{120}$
2. (a) Alfie because he rolled the dice more times than the others.
   (b) $\frac{39}{120}$ (c) $\frac{102}{300}$
3. 44
4. 20
5. 150

75. Risk
1. (a) 0.2   (b) 0.25   (c) 0.25
2. (a) (i) $\frac{28}{4320}$ or 0.0065 (ii) $\frac{45}{6003}$ or 0.0075 (iii) $\frac{19}{3848}$ or 0.0049
   (b) 34
3. (a) £556 010   (b) £84.50

76. Probability trees
1. (a)

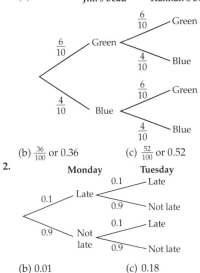

   (b) $\frac{36}{100}$ or 0.36   (c) $\frac{52}{100}$ or 0.52
2.

   Monday   Tuesday
   0.1 → Late: 0.1 Late, 0.9 Not late
   0.9 → Not late: 0.1 Late, 0.9 Not late

   (b) 0.01   (c) 0.18

77. Conditional probability
1. (a)

   First bead   Second bead
   $\frac{6}{10}$ Green: $\frac{5}{9}$ Green, $\frac{4}{9}$ Blue
   $\frac{4}{10}$ Blue: $\frac{6}{9}$ Green, $\frac{3}{9}$ Blue

   (b) $\frac{12}{90}$ (c) $\frac{42}{90}$ (d) $\frac{12}{42}$
2. (a) $\frac{6}{30}$ (b) $\frac{8}{30}$ (c) $\frac{8}{30}$
3. $\frac{96}{336}$

78. Probability formulae
1. (a) $\frac{4}{7}$ (b) $\frac{50}{56}$ (c) $\frac{35}{56}$ (d) $\frac{20}{56}$
2. (a) 0.15   (b) 0.5
   (c) No, as P(A) × P(B) = 0.7 ≠ P(A∪B)
3. (a) $\frac{6}{9}$ (b) $\frac{6}{16}$ (c) $\frac{15}{16}$

79. Simulation
1. (a) Since the figures are percentages out of 100, on time (78%) can be represented by the numbers 00 to 77, late (15%) by the numbers 78 to 92 and very late by 93 to 99.
   (b) On time, late, very late, on time , on time, on time, on time, on time, late, on time, on time, on time
   (c) On one occasion
2. (a) 45–64 gives 20 numbers out of 100, so 4 out of 20.
   (b) V, V, C, S, S, V, C, V, V, V
   (c) In the simulation, vanilla sales were 60% compared to 45% for previous sales.

80. Probability distributions
1. (a) 0.175   (b) 0.525
2. (a) Uniform with all probabilities $\frac{1}{8}$ (b) $\frac{3}{8}$ (c) 2.5
3. (a) Uniform with probability $\frac{1}{10}$ (b) $\frac{7}{10}$ (c) $\frac{2401}{10000}$
4. (a) 0.05   (b) 0.65   (c) 0.455

81. The binomial distribution 1
1. (a) $\frac{4}{9}$ (b) $\frac{4}{9}$
2. (a) 0.01   (b) 0.81
3. (a) $\frac{6}{16}$
   (b) That successive spins of the coin are independent and have the same probability of getting a Head

82. The binomial distribution 2
1. (a) Binomial with probability of blue = $\frac{3}{5}$
   (b) $\frac{150}{125}$
   (c) $\frac{54}{125}$
2. (a) Binomial with $p = \frac{1}{6}$
   (b) $\frac{150}{1296}$
   (c) $\frac{21}{1296}$
3. (a) Binomial with $p = \frac{2}{5}$
   (b) $\frac{720}{3125}$
   (c) $\frac{992}{3125}$

83. The normal distributions
1. (a) 50
   (b) 3
   (c) Test A 1, Test B 1.5, so higher on test B.
2. (a) $\mu = 176$ cm, $\sigma = \frac{14}{3}$ cm   (b) 166.7 cm
3. (a) $\mu = 75$ cm, $\sigma = 1.5$ cm   (b) 380

84. Quality control 1
1. (a) Warning 2.08 cm and 1.92 cm
       Action 2.12 cm and 1.88 cm
   (b) Mean is 2.0425 which is within the warning limits so allow the process to continue.
2. (a) 10.5 mm
   (b) No action after the first 4 samples as they are within the warning limits. Sample 5 fell between the warning and action limits so a second sample should be taken.
   (c) 10.9 mm is above the upper action limit so the process should be halted.

85. Quality control 2
1. (a) The mean is a measure of the average thickness, the range is a measure of the variability of the thickness.
   (b) A second sample should have been taken immediately.
   (c) Point plotted at (5, 0.58)
   (d) The sample range is above the action limit, so stop the process.
2. (a) 30 ml as that is the lower action limit.
   (b) The first 4 samples are within the warning limits so no action required. The 5th is above the upper action limit so the process should be stopped.

Published by Pearson Education Limited, 80 Strand, London, WC2R 0RL.

www.pearsonschoolsandfecolleges.co.uk

Copies of official specifications for all Edexcel qualifications may be found on the website: www.edexcel.com

Text © Pearson Education Limited 2015
Edited by Gordon Davies and Linnet Bruce
Typeset by Tek-Art, West Sussex
Original illustrations © Pearson Education 2015
Cover design by Miriam Sturdee

The right of Navtej Marwaha to be identified as author of this work has been asserted by him in accordance with the Copyright, Designs and Patents Act 1988.

First published 2015

18 17 16 15
10 9 8 7 6 5 4 3 2 1

British Library Cataloguing in Publication Data
A catalogue record for this book is available from the British Library

ISBN 978 1 292098289

Printed in Slovakia by Neografia

Acknowledgements
All images © Pearson Education Limited

Every effort has been made to contact copyright holders of material in this book.
Any omissions will be rectified in subsequent printings if notice is given to the publishers.